Designing for the common good

BIS Publishers
Building Het Sieraad
Postjesweg 1
1057 DT Amsterdam
The Netherlands
T +31 (0)20 515 02 30
F +31 (0)20 515 02 39
bis@bispublishers.com
www.bispublishers.com

ISBN 978 90 6369 408 1

B/SPUBLISHERS

Designing for the common good

A handbook for innovators, designers, and other people.

Kees Dorst
Lucy Kaldor
Lucy Klippan
Rodger Watson
& contributing authors

Part 1
Introduction

Part 2
Projects & reflections

Part 3
Methods

Part 4
Moving ahead

Welcome

This book is for innovators in the public domain as well as designers who want to design for the common good. It tells the story of the design approach to societal innovation used at the Designing Out Crime research centre in Sydney, and its affiliates and friends worldwide.

The wide range of projects you will find in this book use design practices that are built on 50 years of research into how expert designers create new approaches to stuck and sticky problems. The deep knowledge from this body of research has been developed into a model and a method of working that can also help non-designers who want to solve the problems of society in new and original ways.

In Part 1 of this book we introduce what we mean by designing for the common good, and our specific approach to achieving this, called 'frame creation'. The core of the book, Part 2, consists of 21 projects that are diverse exemplars of the breadth and depth of this radical new approach, interspersed with reflections or principles that have emerged from the projects. In Part 3 we dive under the hood of designing for the common good and show 19 of the methods that have proven useful over the years. To round off, we reflect in Part 4 on the meaning of this knowledge for designing our future societies.

Enjoy!

Kees Dorst, Lucy Kaldor, Lucy Klippan & Rodger Watson

Acknowledgements

This book is the culmination of many years of design practice and design research, done in close collaboration with many friends and colleagues. Particular thanks should go to our current colleagues at the Design Innovation research centre in Sydney. And we are grateful for the support the Designing Out Crime centre has received from University of Technology Sydney and Eindhoven University of Technology, as well as from the New South Wales Department of Justice, and the numerous project partners and students that have collaborated with us over the years.

Many thanks are due to all the people who have played significant roles in the projects presented in the book, and who have provided an enduring source of inspiration throughout years of collaboration: Lissa Barnum, Kevin Bradley, Nicole Chojecka, Matt Devine, Paul Ekblom, Jackie Fitzgerald, Lorraine Gamman, Sarah Gibson, Tracey Gwyther, Megan Pee Suat Hoon, Nick Karlovasitis, Fauzi al-Kaylani, Desley Luscombe, Julie McWilliams, Suzie Matthews, Tasman Munro, Ashlyn Park, Peter Poulet, Melanie Rayment, Brad Shepherd, Emmeline Taylor, Brendan Thomas, Adam Thorpe, Amira Vijayanagam, Don Weatherburn and Marcus Willcox, and many, many others. The projects presented here are just the tip of the iceberg, selected out of the approximately 140 projects (and counting) that this community has delivered to date. Curating this selection was by no means easy: there is so much more good stuff out there.

A very special thanks should go to our contributing authors who have been instrumental in the development of this body of work, and have contributed to the writing of the projects: Lindsay Asquith, Mieke van der Bijl-Brouwer, Olga Camacho Duarte, Nick Chapman, Rohan Lulham, Ilse Luyk, Dick Rijken, Rob Ruts, Peik Suyling, Douglas Tomkin, Kim Wan, Vera Winthagen and Jessica Wong. Their contributions are acknowledged as they come up throughout the book. Finally, we would like to acknowledge Jeffrey Kessel and Patrick Forrest for their contributions to the design and layout of this book.

1.

Introd

uction

How, then, shall we live?

The question of how we want to live is before us more or less all the time. That is a relatively recent phenomenon: in traditional societies, the question is hidden behind conventions and rituals that regulate life, and that seem to have done so for as long as anyone can remember. Modern societies have broken free from this mould, only to find that the newfound freedoms and choices come with challenges, and that confusion and suffering are unavoidable consequences of the disintegration of the old structures.

"How, then, shall we live?"* is a difficult question to answer. Societies are complex networks of connections that need to be initiated, developed, nurtured and changed over time. We shape our answer – we build a society – by agreeing on what we have in common, and pursuing what is commonly good for all of us.

In modern societies, this complex and formidable task has been entrusted to 'public sector' organisations, with varying levels of success. Public sector organisations tend to execute this task by focusing on the things that should not be done, by creating rules and laws to regulate public life. Thus they act as a safety valve, putting limitations on people's behaviour to make sure things do not get out of hand. But rules and regulations cannot be more than part of the story: we also need the initiation and creation of new pieces of the societal puzzle. "How, then, shall we live?" is not just a big question, it is a huge creative challenge. To meet this challenge, we need design.

Over the last 15 years, visionary individuals in the public sector have started to explore the possibilities of what design can do. They have worked with designers and artists to create new approaches to societal problems that had proven themselves to be unsolvable by more conventional means. And in the process, these designers and artists have had to re-invent themselves and radically adapt their ways of working to the open, complex, dynamic and networked problems facing their public sector partners. This has been an intense and occasionally challenging process on both sides.

* This question was coined by Steven Kyffin

In this book, we offer what we have learnt so far, through a suite of projects that are presented as first-hand accounts of how design has been used to create solutions to problems in many different societal contexts. Then we will take a good look backstage, listing the methods and techniques from design that have proven to be useful in solving these societal problems. Finally, we will speculate on the shapes that this new design capacity could take in the life of public sector organisations.

Design as an approach

While every professional field has a creative and innovative edge, design is unique in that the creation of novelty and value for people is absolutely central to the profession. As a result, expert designers have developed a treasure trove of sophisticated creative and innovative practices, many of which can also be used outside of the confines of the traditional design domain. So this book is not about design, but it focuses on what can be learned from design by professionals in other fields.

We cannot escape beginning this book with some explanation of the nature of design, as a background for the practices that inform this body of work. The practices themselves will be introduced as briefly as possible, since it is their application to societal problems – as illustrated by the projects – that is far more important.

In this book we speak of design as an approach, rather than as a profession. Hence before we start we need to get rid of some the connotations of the word 'design' that can get in the way of understanding design as an approach. First of all, this is not about designing pretty things. You will see that while aesthetics does play a part in the creation of solutions to societal problems, it seldom leads the way. Secondly, design as an approach is strongly focused on the problem, rather than the solution. At the beginning of a project, we make no assumptions about the nature of its outcome. In the projects you will see that these outcomes range from cultural interventions, to new policies, to the reconfiguration of a public space, to an exhibition, to the creation of a network of stakeholders, and so on. The creation of the solution might involve some conventional design work, but that is not often the case. Finally, this is not about the magic of design. After 50 years of design research we actually know how design works and there is no need to mystify it. In this book we will concentrate on what we can understand about design, and explicate these principles in a way that permits non-designers to make use of them.

Out of the broad range of practices that expert designers have developed over the years, we will concentrate on the practice of reframing problem situations. Expert design is a matter of developing and refining the formulation of a problem in concert with ideas for a solution, in a process called 'co-evolution'. An 'idea' occurs when a bridge is built between the problem and the solution by the identification of a key concept. This means that expert design practices have as much to do with reformulating the problem as with the generation of suitable solutions.

A new approach to a problem is called a 'frame', and the design practice that underpins many of the projects in this book is the process of 'frame creation'. This process has its roots in expert design thinking – the unique, elaborate, multi-layered practices and strategies for framing open, complex, dynamic and networked problems that expert designers have developed. The core frame creation model centres on a nine-step process that addresses complex problems by aligning the interests of the stakeholders related to the problem and concentrating on the emergence of common 'themes' that then lead to 'frames' for action.

The value of frame creation lies in its treatment of a problem situation. Open, complex, dynamic and networked problems often cannot be solved directly, at least not in the terms in which they are presented. The problem and its formulation both have their roots in a specific context that is full of the limiting assumptions that keep us from moving forward. Therefore, the context needs to be critically appraised and altered before the problem itself can be attacked.

So at the core of the frame creation process is a movement of zooming out, then zooming in again. First, the scope of the problem is widened from a consideration of the problem itself and its immediate context, expanding our horizon to the broader field that includes just about anybody who could be interested in the problem or its solution. This jump into the world of speculative thought then allows us to ponder the possibilities that are sparked by the emergence of common themes from within this wider group of stakeholders. An in-depth analysis of the values of these stakeholders in the broader societal field leads to the themes from which new approaches to the problem ('frames') can then be created. The first four steps lay the groundwork, the last steps explore the implications of the potential frames and solution directions for the stakeholders (for more detail see Dorst, 2015 and 'Tunnels & Visions', p.20, and 'Frame Creation', p.162).

Frame creation

1. Archaeology

analyzing the history of the problem owner
& of the initial problem formulation

2. Paradox

analyzing the problem situation:
what makes this hard?

3. Context

analyzing the inner circle of stakeholders

4. Field

exploring the broader societal field

5. Themes

investigating the themes that emerge
in the broader field

6. Frames

create frames by identifying how
these themes can be acted upon

7. Futures

exploring the possible outcomes and value
propositions for the various stakeholders

8. Transformation

investigate the change in stakeholder's
strategies and practices required for
implementation

9. Integration

draw lessons from the new approach
& identify new opportunities within the
network

Societal challenges

The great changes that have faced our societies in the last decades are set to continue. The enormous growth in our technical capacity and communication infrastructure, in particular the internet, has led to the combined trends of globalisation, urbanisation and the creation of a networked society. The pressures that the growing world population and an increase in income inequality have put on the sustainability of life on the planet, the strain on health systems faced with an ageing population, and the need to create a safe living environment in spite of ever scarcer resources have clearly overwhelmed the problem-solving capacities of our government organisations, and indeed of our societies.

Adding to the bewilderment, we have lost belief in the Utopian ideologies that once gave us something to strive for. The Dutch sociologist Hans Boutellier described this predicament as "complexity without direction":

In today's world we have difficulty formulating grand comforting ideas. We hear a cacophony of voices and opinions, see rage and frustration, and observe a lot of ad hoc policy and tentative management… A great deal of tinkering and muddling goes on within politics, educational institutions, the business community…If nobody knows the answer, then we choose what seems to be 'best': best practices, effective interventions, evidence-based strategies…We formulate a politics of risk management and crisis management, of market forces / freedom of choice… We let ourselves be guided by effectiveness and efficiency, demonstrated by performance indicators, supervision and control… (Boutellier, 2013)

It is clear that the public institutions we have created to produce 'common goods' like order and equal opportunities are struggling, and that new approaches are needed. In this book we will explore how design might help. It surely is not the solution to all our problems, but expert designers' practices for overcoming stuckness and failure to progress represent a very good start. Over the last ten years, we have seen 'designers' – people trained in design schools – spreading throughout society and putting their skills to great use. Design is moving away from the making of things to the shaping of processes, policies, organisations and even cities. In the last part of the book, we will be reflecting on the impact that some of these people are having, and the promise that brings for future developments.

Safety as a theme in society

Having said that, to show the variety and depth of what design can bring to a particular societal domain we will need to concentrate. In this book we will look at how design can create a new focus for problems related to safety, as well as a lens through which to examine such problems.

There are some very good reasons to focus on safety, as safety concerns are absolutely fundamental to society functioning at all. Safety is near the base of Maslow's hierarchy of human needs, which means that if safety concerns are not satisfied, nothing else can happen, and the higher-order needs (like love, relationship, self-actualisation) simply cannot be addressed. As Boutellier has explained, the preoccupation with safety and security has become a real force in structuring our society in recent years.

This is potentially highly problematic, for if we really find that we must minimise risk in order to minimise fear, we should be prepared to accept the consequences, which are considerable. Fear is everywhere, and it is the easiest emotion to kindle. A quest for perfect safety leaves little room for cultivating other values; it requires the creation of an immense array of defensive measures to protect us from others and can only lead to total isolation and loneliness. Indeed, total safety gnaws at the very root of our society. It threatens, in a very fundamental way, the existence of anything we could call a 'common good'.

The rising obsession with safety is casting a long shadow ahead. So, we have good reason to see if design practices can help us deal with the valid concerns around safety and security in ways that do not limit freedom and the very life we are seeking to protect.

Beyond the symptoms

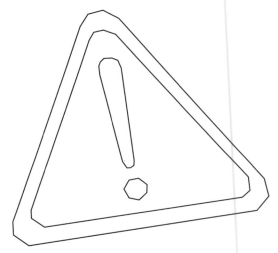

In this book we will use a design perspective to look at crimes and misdemeanors as mere symptoms of things that are wrong in society and see if we can address the root causes in news ways. Examining the issues of safety, security and fear can show us how these deeper underlying issues threaten the common good.

The experience of safety is subjective and hard to pinpoint, and we tend only to realise its importance when things go wrong, when safety is lacking. Boutellier once estimated that more than 50% of all media reports are related directly or indirectly to safety. This daily barrage of messages about threats, accidents, crimes and misdemeanours does influence our behaviour. Beyond reasonable caution lies the abyss of fear. And in everyday life we are defensive, in subtle but omnipresent ways. Double-locking our doors, cautioning children not to talk to strangers, installing alarm systems and cameras, finding ways to spy on our own kids when they go out – the slope can be a slippery one.

Taken to extremes, the imperative to be safe can isolate people, and has its psychopathology in the serious condition of agoraphobia. A dystopia of security might have us stock up on tinned food, buy guns and sandbags, sit at home and wait. The impulse is to champion social cohesion as a panacea, but it usually presents as an alternative extreme and is not very realistic. Common notions of social cohesion seem to be underpinned by an idyllic and ahistorical picture of pre-industrial village life where everyone knows and takes care of each other. But even if such a world ever existed (which is doubtful) our communities now are mostly too big and complex for everyone to be friends in this way.

If these two extremes cannot work, what could be a middle ground? This is a pertinent question, as the two 'solutions', anchored like barnacles at opposite ends of the political spectrum, are the subject of a continuous but largely futile debate.

Conventional problem solving tends to react to such a stalemate by attempting a compromise between these opposing forces and their associated worldviews. But compromise is not the only option. As Caroline Whitbeck (1998) has observed, designers can do something different with this sort of tension. In her book *Ethics in Engineering Practice and Research*, she remarks,

"The initial assumption (within moral philosophy) that a conflict is irresolvable is misguided, because it defeats any attempt to do what design engineers often do so well, namely, to satisfy potentially conflicting considerations simultaneously".

In the projects we will see that we can achieve this by looking very carefully at the problematic situation and then creating solutions that take us away from the initial problem formulation, in another direction entirely. This is where frame creation comes in, and we will see design practices working for the common good.

References

Boutellier, H., (2013) *The Improvising Society: Social Order in a World without Boundaries*, Eleven, The Hague.
Dorst, K. (2015) *Frame Innovation – create new thinking by design*, MIT Press, Cambridge MA
Whitbeck, C., (1998) *Ethics in Engineering Practice and Research*, Cambridge University Press, Cambridge UK.

Projects &

2.

&reflections

Introduction

The projects in this section of the book are a small selection from more than 140 projects undertaken by the team at the Designing Out Crime* research centre in Sydney, and our international collaborators. All projects have been supported by partner organisations (the 'problem owners') who have been seeking a different way to approach the problems they face.

These projects have always served multiple purposes. First and foremost, they have aimed to solve societal problems. Secondly, they have tried to help partner organisations find alternative ways of approaching these problems, and learn something new in the process. And thirdly, every project has helped us develop our practice, and the methods and tools that structure a way of working.

We will now present 21 projects that exemplify the breadth of the problem contexts we have worked in, the versatility of the approach we use and the great variety of solutions that have emerged. Some are products, while others are policies, services and systems, and they are informed by a whole range of discipline areas. These projects are interspersed with reflections – lessons we have learned along the way or discovered by looking backwards. All together these reflections make up a set of principles that we hope will be useful to you.

* The Designing Out Crime research centre was founded in 2008 as a partnership between the New South Wales Department of Justice and the University of Technology Sydney.

Growing up in public

*"Where the night is full of dangers /And the darkness full of fear
And eleven hundred strangers /Live on aspirin and beer."*
– Kenneth Slessor, *Darlinghurst Nights* (1933)

For the best part of a century, Sydney's Kings Cross has been an entertainment district of legendary status. Today, bars, cafes, restaurants, fast-food joints and theatres sit alongside grand art deco apartments, backpacker hostels and the Wayside Chapel, which supports the area's large homeless population. Strip clubs, brothels, a supervised drug injecting centre, a police station, lots of neon and an iconic Coca-Cola billboard complete the physical setting. Edgy, illicit and anonymous, Kings Cross is a place of disadvantage, risk, delight and opportunity.

Tens of thousands of young people come from across the city, the country and the world to let loose on a Friday or Saturday night (or both). The streets are abuzz with people going from bar to bar, and queues for the popular places spill onto the footpath. A trip to 'The Cross' is a rite of passage for eighteen-year-olds celebrating their milestone birthday with large groups of friends. Young men and women, sartorially resplendent, go there to try and meet other young men and women while pretending to be there for other reasons. And, with around 100 entertainment venues crammed into two streets, Kings Cross seems ideally built for having fun.

Yet statistics suggest otherwise. Each weekend night, Kings Cross suffers some of highest rates of antisocial behaviour and violence anywhere in Australia. Recently it was the site of the tragic deaths of two young men who were assaulted by strangers on a night out (see 'Taking Care', p.48). The violent loss of young lives heightened political and public awareness of some of the problems in Kings Cross, fuelling the resourcing of government interventions, as well as panic.

Government responses have included stopping the sale of strong alcohol after midnight, introducing identity scanners for use in 'high-risk' bars and clubs, banning drunks from all venues for 48 hours, establishing a sobering-up centre, replacing glassware with plastic after midnight (so that glasses can't be used as weapons), and introducing strict lock-out laws where entry to bars is forbidden after 1:30am.

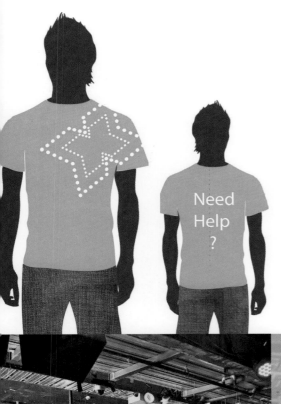

All of these interventions aim to reduce the risk of young, drunk people getting more drunk, fighting and using whatever is at hand to hurt each other. It sounds logical enough, but there are considerable limitations in this response. Many of the restrictions punitively target the physical environment of the clubs and bars, when much of the violence happens out on the streets. Additionally, they penalise *everyone* who goes to Kings Cross, rather than targeting the small number of people who either start out or end up as troublemakers.

At the same time as the state government was working on this risk management response, Designing Out Crime was working on a project with the City of Sydney Council to rethink the overall approach to managing King Cross at night. The team did 'field research' (an educational beer or two) and what they saw wasn't 30,000 delinquents looking for a fight in a cesspool of antisocial and illegal activity, but a very large number of young people trying (and often failing) to enjoy themselves in a fairly unique and exciting place.

The team reformulated the problem from an issue of violence, to a question of how to grow up, have fun, explore and form identity – in public. This reformulation led to a new approach to the problems of Kings Cross.

Thousands of young people converging in one place to have a good time is well-trodden ground: it is the core business of music festivals, dozens of which are held in Sydney each year. Like Kings Cross, music festivals are also at risk of alcohol and crowd-fuelled problems, but festival organisers have evolved a sophisticated discipline of event management that aims to anticipate problems and manage them in a positive, non-antagonistic way.

The team investigated the tried and tested practices of music festival organisers and found a great many that could be applied to Kings Cross, including introducing precinct ambassadors (friendly young people in branded t-shirts); bringing in temporary public toilets and a managed taxi stand; converting laneways into lounge and recreation areas to take pressure off the main streets; installing bright and graphic transport and way-finding signs; programming cultural attractions

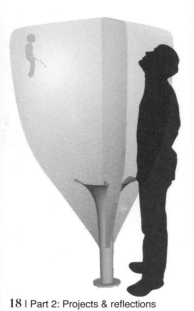

such as public art and urban sports; and creating a 'safe space' where people can go for information, first aid and general help on a night out.

These ideas were well received and many were implemented. But beyond these interventions and the great number of yet-unexplored possibilities, the project helped the City of Sydney to reimagine its own role at Kings Cross. In the words of Suzie Matthews, who was at this time manager of the Safe City team at council: *"We stopped asking how we could prevent alcohol-fuelled violence and started asking: how can we create a vibrant night-time economy?"*

The response to this radical reframing was the establishment of the 'Late Night Economy' team, and the formulation of a comprehensive strategy document, *Open Sydney,* which aims to create a diverse nightlife throughout the city.

Tunnels & visions

The A9 highway around Amsterdam is one of the busiest roads in the Netherlands. To provide for better accessibility to the city, improve air quality and reduce sound levels around one of the bottlenecks of the road, a new 12-lane tunnel will be built, with a park on its roof. The planned construction work will take about 5 years, and these works will impact heavily on the environment – mainly on the adjoining residential neighbourhood, the Bijlmer, a multicultural district of 80,000 people from 186 nationalities.

* Based on a text by Vera Winthagen

A project such as this is a tightrope-balancing act for the stakeholder relationship manager, whose task it is to communicate the program to those impacted, handle complaints and maintain a positive image for the project. The context in which these managers have to operate is one of hard facts and figures: building a new road takes place in a world of strict planning and control, complex process diagrams and tough budgets. Communication with the external stakeholders is professionally handled through extensive consultation processes in order to prevent costly delays. Any delay translates to higher costs.

André Schaminée (consultant at Twynstra Gudde) launched the idea to research the relationship between the construction works and its surroundings in a designerly way. He teamed up with Vera Winthagen (Van Berlo design/ TU Eindhoven) and Tabo Goudswaard (social designer). Together they facilitated a process in which the Department of Public Works and Water Management

(Rijkswaterstaat), the municipality of Amsterdam and the construction company contributed. The process unfolded as illustrated on the right.

After mapping the project and stakeholders, the researchers spent time in the Bijlmer area to glean what was important in the lives and minds of the people, municipality and companies there. This was a very rich process and many fruitful themes were identified. The researchers discovered, for example, that there are many small and excellent entrepreneurs in the area, but that a good many of them are semi-legal, and you have to be part of the community to find them.

This discussion led to the development of a frame that captures the needs of the people and organisations in the area: what if you could reimagine the construction of the tunnel as a new 'temporary economy', and the environment a dynamic stage set? What new connections could be made? With the project framed in this way, the construction workers could be welcomed as temporary inhabitants of the area and supported through small entrepreneurial activity – food stalls, waste removal, childcare, repair services, vocational courses at the local training institutions, and so on – that makes use of and validates the business aptitude in the local community.

The municipality, Department and construction company reacted enthusiastically. The stakeholder relationship managers have been persuaded to look at the area with an open mind, and they see many new possibilities. Most importantly, the framing of the five years of construction work as a welcome time for experimentation and renewal really strikes a chord in the local community. The new activity is a great way to prototype the facilities that can eventually populate the park that will cover the tunnel.

Archeology
All highly professional infrastructure engineering organisations ("planning & control")

Paradox
The need to maintain a good public image despite the impact of the works - while not delaying them

Context
The inner circle of stakeholders: Department of Public Works and Water Management, construction companies, local councils

Field
The various groups of people in the Bijlmer and Gaasperdam, the office workers and commuters, their families

Themes
Talent development, employment, (semi-legal) entrepreneurship, care for the future/next generation, health

Frames
The building works can be seen as a 'temporary economy', leading to welcoming the workers as temporary inhabitants, supporting them with services that can then become permanent
Seeing the construction site as a dynamic stage set

Futures
Mapping the existing food offerings in the area, creating bespoke food stalls, setting up childcare services, introducing new courses at the local vocational training center, setting up local firms to deal with the waste materials of the building works, etc

Transformation
Support bottom-up initiatives from the local community (instead of 'planning & control')

Integration
Learnings: how to foster the new opportunities that arise for all parties when a big building project comes to a neighbourhood, and how a temporary situation can be used to experiment with the permanent inhabitation of the space once the works have finished

Survival of the fittest

Clothing is a signature product for department stores, whose brands are built on the promise of the latest fashion trends and the thrill of immediate personal reinvention. Clothing is also a highly coveted item for thieves.

At one of the largest Australian department store chains, tens of millions of dollars worth of clothing is being stolen every year. The company's security managers have discovered that people engaging in occasional, opportunistic offending caused the vast majority of losses – more even than the combined efforts of repeat career shoplifters – and that most theft occurs in the fitting rooms, the only place in the store offering sufficient privacy to conceal unpaid-for goods. The department store's loss prevention team engaged Designing Out Crime to explore new ways of preventing shoplifting from fitting rooms in their stores.

* Based on a text by Rohan Lulham

There were two clear and immediate challenges with redesigning fitting rooms with the explicit aim of reducing loss. The first: the company's loss prevention practices and procedures were already regarded internationally as being close to best practice. How could a team of students improve on industry-leading expertise? The second challenge was an issue of organisational tension centred on the retailer's imperative to sell – at all costs. Customers make important sales decisions while trying on clothes in fitting rooms, and decisions to purchase are influenced by relatively intangible aspects of customer experience, such as how comfortable and relaxed a customer feels. Anti-shoplifting measures in fitting rooms aim to reduce the extent to which would-be shoplifters can conceal and be concealed (e.g. by taking away the shelf and the chair, by creating large gap under the door or curtain for surveillance, by limiting the size of the fitting room, &c.) But in so doing they also tend to destroy any sense of the delight and luxury that legitimate shoppers expect from department store shopping, dampening customer experience and – critically – directly

reducing sales. The company's marketing and sales team was rightly suspicious of the loss prevention approach because of its demonstrably negative impact on sales. And in a sales business, marketing trumps security every time, leaving the loss prevention and security team somewhat impotent to do its job.

Seeing that loss prevention was stuck, the Designing Out Crime team researched the broader context of department store retail to find a way into the problem. They discovered that the greatest threat to bricks-and-mortar department stores – even greater than the threat of lost sales through theft – is competition from an increasingly sophisticated online marketplace and changing consumer demand. They heard from the sales team how bricks-and-mortar retail is under pressure to reinvent itself, to find new ways to create an evocative, enticing and memorable experience.

In this context, the fitting room experience emerged as just about the only advantage that bricks-and-mortar stores retain over online retail – the opportunity to engage a customer's senses (touch, smell, feel) in the ritual pleasure of trying on clothes. The Designing Out Crime team created a productive frame to reimagine the fitting room as a place where customers would choose to spend time, almost as a shopping destination in its own right.

The fitting rooms, which have been literally peripheral in department store layout as well as symbolically peripheral in considerations of customer experience, were redesigned to be prominent both within the store and the shopping experience: centrally located, visually unmissable, and suggesting activity and excitement, something like a catwalk. Repositioning fitting rooms in this way also reduces opportunity for theft by increasing natural surveillance and heightening the sense of risk of being caught.

Additionally, the fitting rooms were designed to create a seamless connection between customers' in-store experiences to their online world. An interactive stylist console in the fitting room space allocates fitting rooms (through barcode scanning), suggests complementary products (e.g. hats, accessories) and allows customers to register a loyalty card that 'remembers' their favourite items, gives them advice based on previous purchases, and notifies them when their favourite items go on sale.

These concepts were well received by the loss prevention team and helped them to become a positive voice for change in the organisation. The department store has recently launched a new retail concept which brings many of the store's customer services to the centre of the floor. This new approach, which is customer-centric in a very literal sense, makes the most of the sensory experience of shopping that can only be had in a bricks-and-mortar store.

please proceed to the selected fitting room

OK   CANCEL

What we want more of

There are not many strict rules and principles in designing for the common good – the projects presented in this book may even seem to have little in common. Each is different in subject, size and scope and the approach taken depends very much on the concrete problem situation.

But if there is one rule by which we always abide, it is to design our way out of a problematic situation by concentrating on 'what we want more of'.

If a public space is prone to crime, for example, one could always choose to take defensive measures to stop the bad things from happening. In our philosophy, such defensive measures – which are often literally fences and barricades – are not the preferred option since they limit freedom for all of us and detract from the common good. We adamantly choose to go in the opposite direction, making the space more usable for the good people, knowing that this will crowd out crime. As we will see later, carelessly designed public spaces (or spaces that are not designed at all) can often be redesigned to support activities that fill the space with positive activity. The existence of deliberately and thoughtfully designed features in a built environment sends the signal that the place is cared for, which changes the way everybody feels about and uses it. The simplest litmus test for the quality of a public space is always "would you take your kids here?" This test can be adapted and applied to all projects, whether they are about environments, systems, processes, or policy.

This principle applies not just to the design of public spaces, but aiming for 'what we want more of' is for us a mindset, applicable across all projects. In these first three projects the reader has already seen this principle at work.

But wait!— THERE'S MORE!

Be magnetic

A problem with designing for the common good is that "the devil has all the good music": it seems devilishly easy to attract people by playing on their human weaknesses (egotism, greed, fear, aggression), while the values we strive for in designing for the common good may all seem a bit unspectacular, 'soft' and do-goody.

The fact that these values are not directly attractive in and of themselves means we need to be strategic in how to achieve them. The first step in this strategy is to understand what competing attractors there are in the problem situation, and what are the deeper values behind this attraction. Often these values can be expressed in many different ways – ways that would be disastrous, as well as ways that are beneficial for the common good.

Take, for instance, the stereotypical situation of a group of young men on an evening out, and the need to establish the 'pecking order' within the group as a way of expressing identity and manhood. Actual attempts to establish a position in the group can manifest in boisterous behaviour and fighting, which is a particular problem in the public domain where innocent people might get hurt. This is one of many elements behind the violence in Kings Cross (see 'Growing Up in Public').

Searching for an alternative pattern of behaviour, one could look at how these conflicts are not regulated or suppressed, but rather ritualised in tribal societies. If we looked at violence between young men as ritualised conflict, there might be a solution in providing urban games or sports so that various members of the group can display their prowess and be recognised for it, without hurting anybody. In this way, we can redesign the situation to move the actions in the right direction.

Gone shopping

The two largest supermarket chains in Australia each lose millions of dollars annually through shoplifting. Losses have to be covered somewhere, and are borne by the non-thieving consumer through increased prices.

Traditional methods of reducing in-store theft include camera surveillance (CCTV), plain-clothes security, staff training to spot offenders, electronic tagging of highly desirable items and placing expensive or highly desirable products in secure cabinets that are accessible only to staff. Security and loss-prevention technology continues to increase in sophistication, yet in the supermarket context it stumbles in one crucial respect.

+ Based on a piece by Douglas Tomkin and Jessica Wong
* *Gone Shopping* is the title of a book on shoplifting by Lorraine Gamman

Retail sales are made through proximity and connection between a would-be consumer and a product. The easier it is for a customer to personally examine an item for sale, the more likely a sale will result; accordingly, every effort is made by the retailer to make products accessible to shoppers. At the same time, a product within reach is easier to steal than a product under lock and key. The classic loss-prevention dilemma is how to reduce opportunity for theft while encouraging sales, when both sales and theft are enabled by the same mechanism.

Having been asked to look at this problem by a leading supermarket retailer, Designing Out Crime undertook an audit of frequently stolen items, the stores most susceptible and the modus operandi of thieves. While there is great variation among the products that are commonly thieved – everything from fresh meat to electronic goods – the most vulnerable products are small, expensive and suitable for resale online. Retailers have reported that stolen batteries, razors and cosmetics can appear on e-commerce sites within an hour of leaving the store. CCTV footage has

revealed shelf-sweeping to be the most popular tactic used by offenders. The thief stands with their back to the CCTV camera and with a quick movement of the arm 'sweeps' dozens of products at once into an open bag lined with foil (to avoid tag identification at the exit). Shelf-sweeping wreaks havoc on stock replenishment systems because stolen items do not registered as having 'gone' (since they have not been sold), meaning that the product stolen is not available to legitimate customers until a staff member finds it missing from the shelf.

The traditional role of store security is to reduce theft by warning potential offenders of the likelihood of getting caught (via signs and CCTV) and apprehending those who try. This defensive approach may catch crooks but can also counteract sales goals. All things considered, it can be tempting for retailers to turn a blind eye to theft. Clearly this problem is very stuck.

The approach adopted by Designing Out Crime sought to frame the crime problem from a positive standpoint, and to theme solutions around sales objectives. In the retail context, sales priorities heavily outweigh security concerns; thus, solutions that encourage sales are more likely to be adopted than those focused squarely on loss prevention. The Designing Out Crime anti-theft shelving design satisfies both aims.

Features of the shelf include a product information strip (engaging customers to increase sales); a transparent flap that needs to be lifted with one hand before a product can be removed (requiring the use of both hands and making shelf-sweeping impossible), and a light that both illuminates the product and gives the signal that sensors have been triggered.

Given positive customer feedback and early indications that theft decreased through use of the shelf, the supermarket initiated the development of a second version of the shelf to undergo more detailed and prolonged assessment. This project created a new product that wasn't a simple compromise between the two opposing forces of loss prevention and sales, but shows how both can be seamlessly incorporated into a single design.

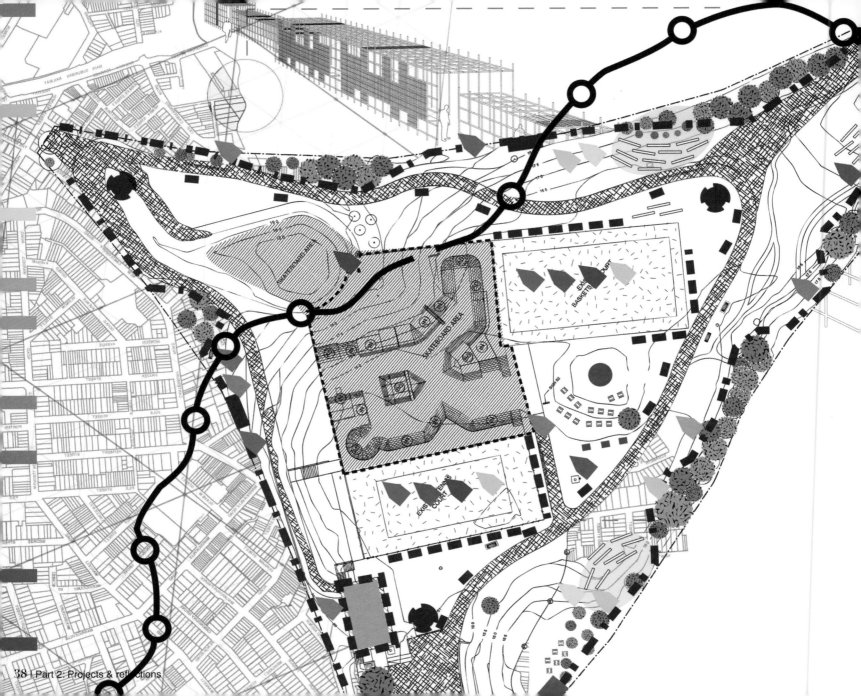

Are we there yet?

The announcement of a new rail line in the inner suburbs of Sydney was a good news story for all.

The community had lobbied for better public transport for decades and a 12km strip of disused freight train track that rambled through residential suburbs to the city provided the perfect opportunity for conversion into passenger rail.

The government was confident that the line would be very popular with city commuters. However they feared that some of the stops might be dangerous, especially at night. Two in particular were quite isolated, so Designing Out Crime was invited to give advice on how to make them safer.

* Based on a text by Nick Chapman

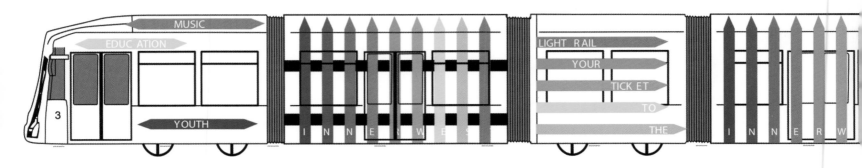

The two stops were quite unique, with very different issues and opportunities. One was in a desolate, semi-occupied industrial site earmarked for residential development. Passengers using this stop would be required to walk along crumbling footpath down a dark, pot-holed lane lined with shuttered doors and high walls. The other stop – the last on the line – was accessible only via a triangular park surrounded on all three sides by railroad tracks. Shabby and dilapidated infrastructure, including a skate bowl, tennis court and badly grafittied toilet block, made the park a threatening pedestrian environment, particularly at night. Poor lighting completed the effect. The park was perceived to be a haunt for drug taking.

The project produced many feasible and attractive designs for improving the look, feel and safety of the two sites. But Designing Out Crime wasn't sure that isolated interventions would go far enough towards making the line safer, and the project team decided to take on a much broader interpretation of the problem. If people perceived that the line wasn't safe, they'd be less likely to use it, which would in turn make it less safe. So the challenge was not just to make two stops safer at night, but to promote use of the line.

The line links suburbs known for their unique cultural heritage. Over many decades, immigrants from Italy, Greece, Portugal, Vietnam, China, and Korea have established cafes, restaurants and specialty supermarkets in these areas that cater to their own communities and culinary traditions. As it shuttles to and from the city, the rail line gives passengers both views of and easy access to places that offer new experiences. One recommendation, thus, is to promote these new connections through a curated suburban food safari, with participating restaurants becoming part of the branding of the line.

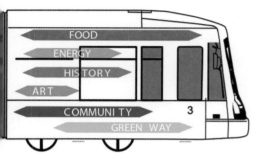

The line also happens to pass a waterside park that is a very popular place to exercise, and a newly created nature walk that runs alongside the line. The team suggested bringing these three elements together with the rail line becoming the backbone of an exercise and a nature education trail. In this way, a project that was initially about safety culminated in an array of ideas to take people on journeys – of cultural and culinary delight, a healthy active life, and the discovery of nature in the urban jungle.

PARKLAND

RETAIL

COMMERCIAL

RESIDENTIAL

MIXED USE

INDUSTRIAL

Two problems at once

When a problem is so confounding that smart people haven't managed to resolve it yet, one approach is to find another problem, add it to the first and see what you come up with.

Sometimes, the reason an organisation or community hasn't solved a problem is that any solution it conceives of directly contradicts its dearest-held (if little-voiced) values or jeopardises its fundamental priorities or assumptions. In 'Gone Shopping' (p.34), insight into the corporate state of mind showed how fears about shrinking profit margins due to theft were considerably outweighed by concerns about harming sales; and that theft-prevention solutions (such as a heavy security presence, or locking products in cabinets) had not succeeded because they were viewed as an unacceptable threat to sales. In this project, addressing the need to sell in tandem with problem of theft catalysed solutions for both.

In the project 'Are We There Yet?' (p.38), the design concepts sought to address the problem that was initially presented – keeping rail commuters safe at isolated stations – by concentrating on the more pressing issue of how to attract people to a new train line.

So when a problem appears hard to solve, another lurking, insidious problem might be its unexpected saviour. Often in our work, the second problem comes to light because there was something just a little bit fishy about how the first problem was presented or constructed, and it prompted us to ask: what else is plaguing the organisation, company or community? What else is sapping people's time, or keeping them awake at night?

Who's afraid of Stratums Eind?

In the city centre of Eindhoven is a long street lined with bars and pubs, called Stratums Eind. This street is the hub of nightlife in the city.

A couple of years ago, the city of Eindhoven raised the question of how to increase safety for the teenage girls who go there on Saturday nights. According to the city council, the safety image of Eindhoven was suffering from 'the current situation at Stratums Eind'. Unsure of precisely what this situation was, the team from Designing Out Crime began to investigate.

Curiously, analysis of police data revealed hardly any safety-related incidents involving girls at Stratums Eind; reports of violence and abuse mainly involved male victims. And initial field research revealed that the girls actually felt safe there. Naturally, then, the next question was: who is afraid of what, and why?

* Based on a text by Ilse Luyk

The City of Eindhoven, clearly, was afraid of gaining a poor reputation for safety among its constituents. This fear was well founded, since it turned out that parents were very concerned about their daughters' safety. Discussions with the parents revealed much about their fears. Most of the girls in the research group lived in the quiet suburbs of Eindhoven. Their parents rarely ventured into the city centre themselves, and their main source of information about what was happening in town was the local newspaper. And whether deliberately or accidentally (mostly the former), the teenage daughters tended to keep their parents in the dark about what happens on their nights out, and rarely came home when they said they would. Totally alienated from the city nightlife experience (except for occasional, negative media reports) and lacking information about their daughters' whereabouts, parents suffered in a state of unremitting anxiety – in spite of lack of evidence of actual danger for teenage girls at Stratums Eind.

And what of the daughters' fears? The daughters were unafraid of Stratums Eind, but they *were* apprehensive about cycling home alone to the suburbs, late at night. Where possible they would cycle in groups with friends – certainly, this is what they would promise their parents.

So what began as a public safety problem ended up as largely a problem of lack of information, misperception and fear.

The first solutions were directed at parents – or, rather, at improving Eindhoven city's image by allaying parental concerns. The city of Eindhoven invested in guided tours of Stratums Eind where parents could learn about safety measures, crime statistics and alcohol policy. They were also invited to visit some popular cafés at Stratums Eind, to experience the nightlife for themselves.

The second solution aimed to solve the problem of parents not knowing their daughters' whereabouts with a non-intrusive communication tool in the form of a bicycle key. The key remotely communicates the status of the bike and its rider (the statuses are: cycling – arrived at destination – home) via a simple LED-light that changes colour according to the status.

For the girls cycling home, Designing Out Crime developed a smartphone app that makes it easy to form cycling-home groups from among their own social networks. Another concept targets the nighttime cycle path that connects the Eindhoven city centre with the main suburbs. These paths would be dynamically lit with LED lights to show a 'green zone' – a moving zone of green lights indicating the pace that cyclists should follow in order to catch a green light at the next set of traffic lights. The result of these moving green zones is the clustering of cyclists (for group safety), less running of red lights and less stopping. These measures help them to get home safely.

Taking care

The Thomas Kelly Youth Foundation (TKYF) was established by the grieving family of 18-year-old Thomas Kelly, who died after a single fatal punch from a stranger at around 10pm on the 7th of July 2012. The teenaged perpetrator had been drinking heavily; the assault occurred in Kings Cross, Sydney's most popular nightspot.

The random killing of one young man by another garnered public outrage and empathy, as well as massive media attention. Long-running public campaigns to curb problematic drinking were given renewed vigour, and the Kelly family was at the centre of community demands for tougher restrictions on the sale of alcohol, as well as tough sentences for offenders. In the midst of all this, they launched TKYF with a much broader vision for 'a new approach to engage with young people and their families and reduce the level of harm in our cities.'

TKYF approached Designing Out Crime and over a series of conversations and workshops we worked with the foundation to disentangle and clarify some of the notions in their mission. What would a new approach look like? How could the foundation engage effectively with young people; and what, precisely, are the harms we want to reduce? How can we reduce harm without curtailing fun and infringing on freedom?

Public discussion about youth is habitually fear-driven. Doubt and concern about 'young people today' – their interests and choices – is a seemingly perpetual theme in social discourse. Government policy and public debate about youth and harm is preoccupied with alcohol and its association with violence. Drinking too much is an acknowledged social problem, for young and old, and there is political gain in being seen to be spearheading preventative or punitive measures to reduce violence and problematic drinking, particularly among youth.

But being young is not a health hazard, however the media might portray it. It is an important life stage through which we hope each individual passes safely and happily, and alcohol-related danger is but one of many factors that can disrupt the safe journey of a young person towards adulthood. Most adults recall a time in youth when their

interests seemed in conflict with those of their parents; older generations in any era will attest that they have trouble understanding the behaviours, interests and choices of younger people. In the family or community context, this lack of understanding begets conflict and at a political level, it equals inadequate responses and misallocated resources.

The outcome of the discussions with TKYF was a proposal for the establishment of a 10-year program that involves young people in the creation of initiatives to help them grow up safely, while drawing on expert knowledge from diverse academic and practical fields. The mission of the program would be to redefine the way in which community deals with the complexities of life for young people.

The program would take a broader view of what safety means to (and for) young people. It investigates the recurrent, as well as the current themes and realities of youth. What does it mean to be young, today and forever? What is going on in the lives of young people, and society more broadly, that propels them along one course or another? Where do young people get their information and how do they make decisions? How can unsafe habits or behaviours be interrupted? In short, what are the mechanisms by which we can create safer society with and for our youth?

Radical generosity

Some years ago we were approached by the local council that is responsible for an area of Sydney that includes 'The Gap' – a spectacular cliff near the entrance to Sydney Harbour that unfortunately has a reputation in the city as a place where people go to suicide. The problem that the council presented to us was basically that Don Ritchie was getting too old to walk his dog at twilight...

Don Ritchie lived in one of the last houses next to the nature reserve that includes The Gap. He used to walk his dog at twilight, and approach people who were looking distressed, asking them "Why don't you come in for a cup of tea?" Over the years he chatted with hundreds of people at his kitchen table, and saved many lives (the estimate is about 400).

Now that Don was getting on in years, what should the authorities do? They realised that putting an emergency phone at the spot was probably not going to do the trick – too deliberate, too institutional, too obvious, too authoritarian and, despite good intentions, also too inhumane. The fact is that Don's kind deeds, emanating from basic human goodness, cannot be organised at all without losing their essential power.

Human goodness and generosity are incredibly powerful forces in our lives. All the real quality of the projects presented in this book comes not from the cleverness of the methods used or the funky visualisations. Real quality comes from the fact that somewhere along the way, somebody involved in the project (a designer, a local council member, a social worker, a student – whoever) was personally inspired to bring his or her own force to bear on the problem.

The product is a process

Often the outcomes of our projects are not products in the conventional sense of the word, but processs, policies or organisations. In engaging with our partners and collaborators in the problem solving process, there have been a few occasions where the suggested solution goes beyond the capability and responsibility of any existing organisation, and would require the creation of a completely new organisation in order to implement the solution.

This in itself shouldn't be a surprise. In the private sector, new businesses 'start-up' all of the time. Often these businesses fill a gap they have found with an idea they have created. Or they create an idea in one context and see an opportunity to market it in another. Businesses succeed or fail on the value that their idea creates and on their business skills.

A start-up in the public sector is a little less common. Designing Out Crime is an example of a start-up. The government saw a need that wasn't being met and asked universities to respond to that need. It is fitting, then, that in our work we sometimes come across situations where the outcome of a project is a new organisation.

The example of the proposal from the 'Take Care' project (p.48) is a case in point. The solution in this project wasn't a new intervention to the problem of alcohol-related violence. It was a whole infrastructure that would be set up to support the sector to find new ways of working, and to develop new interventions that cut across organisational silos.

Serenity and security

The luminous white shells that form the roof of the Sydney Opera House are among the most recognisable and most photographed architectural structures in the world. Perched on a peninsula between the sea and the sky, the Opera House is unbelievably serene and peaceful. It can be (and is) seen from multiple vantage points in Sydney, by locals and millions of international visitors every year.

What better canvas, then, for an anti-war demonstration? In 2003, activists scaled the large western-facing shell and painted the slogan 'No War' in red paint. It wasn't the first, nor the last time that people climbed the shells illegally; in 2011, climbers hoisted themselves and a banner to protest against commercial logging. Attempts at scaling the shells have been surprisingly frequent, and a determined climber can find multiple access points relatively easily. But the risks posed by an incursion of this sort are considerable, and could damage the UNESCO world-heritage listed building.

* Based on a text by Douglas Tomkin

The Opera House Trust is investigating ways to prevent security breaches on the shells. Responses have included fencing off vulnerable spaces between the shells and increasing security presence. But the fences both spoil the aesthetic of the building and hinder visitors from wandering between the shells to experience fully the magnificent building and its waterscape. There has to be a better solution.

The design challenge began with a comprehensive briefing from the Opera House management and police covering security, heritage, planning, events, maintenance and the bureaucratic procedures governing change, as well as architect Jørn Utzon's design principles that guide any possible changes to the building.

Going through the frame creation process, the team broadened the context to include many stakeholders in the city. This search for a new angle to approach the safety issues led to identifying two underlying principles: 'connecting' and 'sensing'. 'Connecting', because although Sydney people are duly proud of their Opera House, those who are not regular patrons actually feel little connection to it. This, the team found, was a result of the fact that the

Opera House, swarming with millions of tourists year-round, lacks public activities that make you want to come back to it and renew your acquaintance. And unfortunately, the parts of the public space around the building that are now closed off for security reasons would exactly be the places where one could organise small events and create the sense of the intimacy required for locals to develop and sustain a 'love affair' with this special place. The theme of 'sensing' referenced the idea that the senses that should be delighted by the Opera House. It is a spectacular spot to experience the elements (sun, water, wind, silence) and achieve the 'spiritual uplift' that the architect envisaged.

This led to many ideas on how to help locals cultivate closer ties with the Opera House. These include a volunteer team of cheerful, local ambassadors to provide information and direct visitors; and curated events on the forecourt that cater to locals, such as yoga at sunrise.

One could try to create a sense of dematerialised lightness, or weightlessness, in tune with Jørn Utzon's original sketches for the Opera House (in which the white shells have an open, cloudlike structure). Similar projections on the floor inside the foyer and on the paving outside can be used to blur the boundary between inside and out, visually creating the sense of lightness the architect envisaged in the original drawings. Thus, the podium could be a fluid landscape, which locals return to often to drink in the new experiences.

The infrastructure of sensors and lights that supports such events also doubles as a subtle security measure. These would increase the number of 'eyes on the shells', while giving Sydney people the chance to re-experience and feel at home at the Opera House.

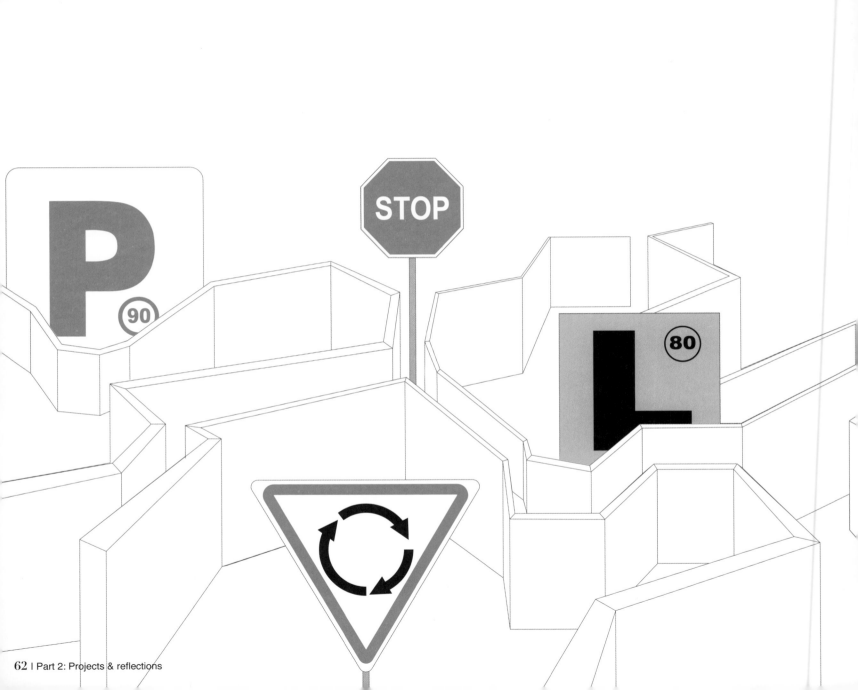

License to drive

Every year in Australia, a number of people go to prison for being caught driving without a license, and not paying the ensuing fines. At first glance this may seem fair enough –after all, driving without a license implies driving without ability or skill, which sounds like a dangerous thing to do. But scratch the surface, just a little, and it becomes clear that people are going to prison not because they can't drive (they can!), but because getting a license is prohibitively expensive and requires navigating an impenetrable and hostile bureaucracy.

In the remote areas of Australia it is possible to drive for hours without seeing another vehicle. These are places where public transport is non-existent and essential services (food, health) can be hundreds of kilometres away. The people who live here – mostly Aboriginal people – learn to drive however they can, in whatever vehicle is available, because they just have no alternative.

In 2013 this issue was selected as a case study for a government educational program to reframe a societal problem. The challenge was to reduce the significant number of Aboriginal people going to prison for unlicensed driving. The team began by mapping the process that a person must follow to obtain a driver's license. Major milestones were identified, as were the assets and skills required to pass each milestone, such as money, access to a registered car, a licensed adult to be a teacher in the first stage of supervised driving, English literacy to take the tests, and so on.

Many of these hurdles are clearly insurmountable obstacles in these remote contexts. Next, interview data – anecdotes from Aboriginal people with experience in driving without a license and the ensuing court processes - were overlaid onto a large process map, highlighting the obstacles and

Getting a license

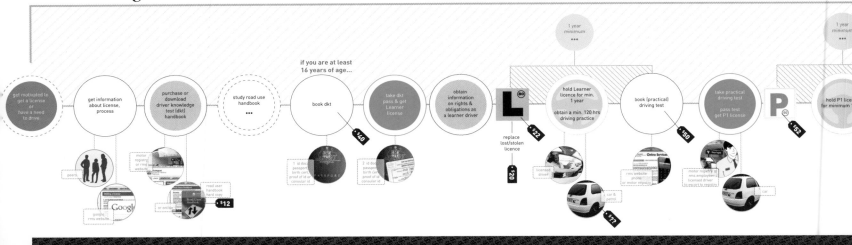

BARRIER:
Unlicensed driving is the norm: feel they can avoid detection

'Yeah, I just take the back roads, plus the cops are like clockwork. I drive when I think they aren't patrolling'

female, 22 years, regional

BARRIER:
Feeling uncomfortable at RTA motor registry

'Having a black fella coming out would make it a lot easier and that and they explain it good'

male, 52 years, remote

BARRIER:
Financial barrier to afford multiple tests

'They used to have a program on that RTA computer for people to test their knowledge and stuff. But no-one uses it because they think it's a waste of time and money because they can't afford to go through the L's and green P's and red P's'

male, 39 years, urban

pain points for the students to gather around and talk. This revealed that the licensing process is incredibly long and costly, and while it is well engineered to test compliance with administrative procedures, it is poorly designed to test a person's true driving skill. What's more, the perverse situation where a parent is consistently fined for unlicensed driving while bringing her children to school, or bringing an elderly community member to the doctor, is intolerable.

The stream of solution ideas that followed moved away from focusing on improving the bureaucratic procedures in favour of helping these communities to meet their transport needs (a car and driver sharing system, for example). Small but significant projects like this – in which a government designs for people - will help to create a more equal society where people aren't unnecessarily imprisoned.

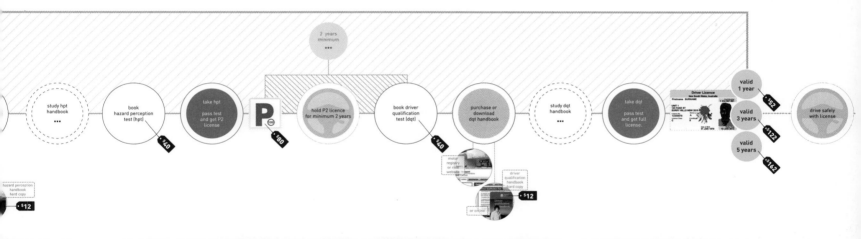

BARRIER:
Literacy difficulties

'I'd get one [a license] but I can't read. I can drive good but that test is shit.'

male, 17 years, urban

BARRIER:
Lack of registered vehicles for practice

'Yeah right! How many Kooris you know who got a license and will let you drive it. Unless you drive them home when they're pissed you never get a run'

male, 17 years, urban

BARRIER:
Lack of registered vehicles for practice

'I'd say there's only three people with a licence here in this community. And two of them are on their P's'

female, 36 years, remote

BARRIER:
Literacy difficulties

'I can't read or nothing. Talk to me. If you tell me I'll know but I'm not reading this crap.'

male, 17 years, urban

Are you being served?

People filling up at a petrol station (called a 'service station' in Australia) and leaving without paying (a 'drive-off') costs the industry $66 million annually. Incremental increases in the price of petrol correlate with increases in the theft of petrol. Petrol theft reduces revenue and profit for a petrol station, and also costs valuable time to report, investigate, and prosecute.

The solutions suggested by police, such as CCTV cameras and pre-payment at the pump, were based on the objectives of making it harder for people to steal, or easier for them to be caught. The rationale for both was sound but neither was practical. CCTV cameras don't necessarily deter drive-offs, and thieves obscuring license plates limit their usefulness. Mandating pre-payment might stop theft but runs contrary to the business model of petrol stations, which relies heavily on customers buying other things – chips, drinks, newspapers – in-store. Early trials of pre-payment at high-theft locations actually harmed in-store sales, costing petrol stations more than the thefts.

* Based on a text by Lindsay Asquith

This project aimed at breaking the impasse between police and petrol stations. Research showed that while drive-offs were a significant problem, the petrol station industry had more preoccupying (and expensive) problems. In addition to being targeted for theft, petrol stations are also targets for armed hold-up. This crime situation is rare but incredibly traumatic for staff; often employees cannot return to work or choose to leave after witnessing an armed hold-up situation. Low salaries and limited prospects for career progression mean that satisfaction among staff is low and turnover is high, which is very costly for station owners.

Petrol is regarded as a necessity, and not something that a petrol station should have to go to any trouble to sell. However, it's not the petrol that makes the company's profit, but the shelves of packaged products that motorists walk past on their way to pay. Viewed as a retail organisation – albeit one that does not typically

Hi!
Please
proceed to
PUMP 4

excel at retail – customer experience, store environment and product offerings become more important. The oxymoronically named 'service station' offers an underwhelming and even unpleasant customer experience with little or no real service for customers. Customer loyalty is virtually non-existent, and most sales are impersonal, grudging transactions, which is not good for staff morale. The products on petrol station shelves are expensive and generic and rarely exactly the thing you're looking for, since stations must aim to cater for an impractically large variety of potential customers.

Designing Out Crime focussed on the theme of 'service', and came up with the 'My Servo' concept. My Servo appeals to the contemporary expectation of service being customised or tailored (through technology), as well as the traditional notion of retail service as direct, skilled help from a human being.

My Servo aims to solve both customer and staff dissatisfaction directly, creating opportunities for service station employees and customers to get to know each other, to improve sales and reduce the likelihood of a service station being targeted for

petrol drive-offs – which is increased if the service station is perceived as 'faceless'. A membership and rewards program (activated via swiping a card before using the pump) would improve customer loyalty and give station owners data they could use to improve, for example, the products they stock. Physical changes to service station layout – made under the pretense of improving customer service – could help target drive-offs; for example, high-theft lanes or pumps would be reserved for regular customers with a card.

Design can be evil

Design is a humble servant that creates tools (products, services, whole systems) that help people to attain their goals and satisfy their needs. The designer, then, is in a powerful position, since the goals and needs of people are shaped by the tools they have at hand to achieve them. So the influence of design comes with great responsibility and is also open to abuse: by providing certain tools, designers have the power to influence the behaviour of people without the 'victims' realising so.

For instance, fast food chains lure you in with seating arrangements that look nice, while the chairs themselves are made uncomfortable on purpose. This is to make you leave within 10 minutes and vacate the spot for the next unsuspecting victim – pure manipulation.

Sometimes it can be tempting to take a shortcut that creates the impression of quickly resolving a problem. However if we provide a solution that creates problems for someone else, then it isn't designing for the common good.

It is important always to 'do no harm', and anchor back to an acute understanding of the 'common good.' We have the obligation to protect unsuspecting users from abuse, and from ourselves.

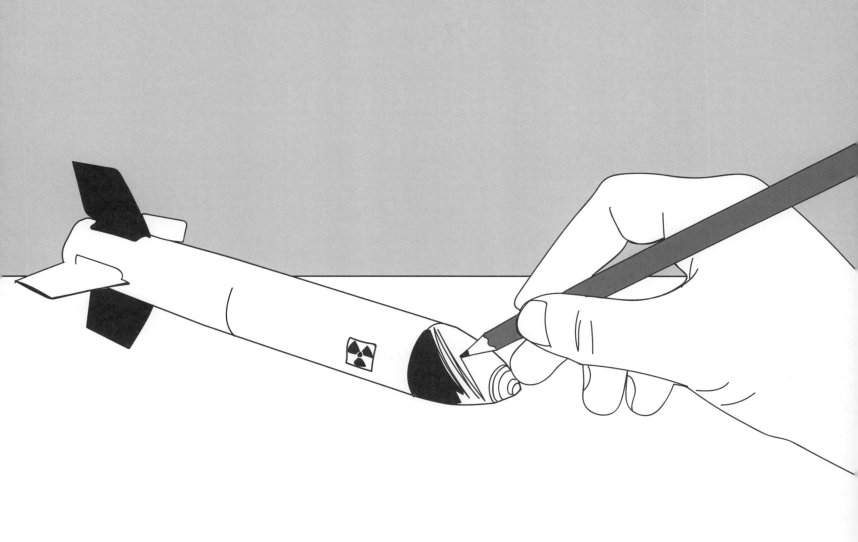

Problems aren't what they seem

We are acutely aware that there is always a danger that we might be solving the wrong problem. If a problem is set from an old or narrow perspective, solving it from that position will probably not help at all.

This principle is the starting point for many, if not all, of the projects presented in this book. The problem itself needs to be reformulated (reframed) before we can solve it. As we saw in the project 'Are You Being Served?' (p.66), there are many simple ways of stopping the theft of petrol, like mandating a pre-paid system. Such a system could even be designed to be a bit elegant, like the now-ubiquitous public transport smartcard, requiring little forethought or planning from customers.

But a solution like this – one that solved the single, original problem of petrol theft – would never be acceptable to a company selling petrol, because in spite of appearances their business models rely not on petrol sales, but on hungry, thirsty, bored car travellers walking into the shop and succumbing to brightly-packaged snack food and magazines. However, by investigating and reframing a problem to take in the petrol company's other troubles– such as high staff attrition – more interesting and versatile solutions immediately become clear.

The fact that problems are often not what they seem is a source both of great worry and great inspiration when designing for the common good.

Breaking the deadlock

Generally, people mean well. But within organisations, sometimes whole teams or departments are at odds, and the actions of one can make it hard or impossible for the others to achieve their aims. Then, whole ecosystems evolve around a problem in a way that maintains the problem, rather than resolves it.

It is commonplace to go on doing what you are doing because that's what your job/team/department was established to do. But in 'To Protect and Serve' (p.82), we see an example of how focusing on victim needs provided a much-needed reset button that helped the department transform the justice system.

Another example is seen in the project 'Serenity and Security' (p.56). The heritage department has been set up to keep the building as it is; its designation as a UNESCO heritage site more or less turns the whole site into a museum piece that is frozen in time. But at the same time, it is a functioning concert venue that will continually need to be updated to keep abreast of new technologies. And problems like trespassers on the roof inevitably require changes to the building, but heritage restrictions prevent it.

This deadlock has cleverly been resolved by commissioning the formulation of the 'Utzon rules', in which the original architect of the Sydney Opera House explicitly stated his design intentions. This enables much-needed changes to the building to happen, as long as they are in line with these intentions.

Putting two and 2 together

When trying to make sense of a problem situation, relying on a single data source can easily lead to a superficial understanding of a problem. This will likely lead to a solution that treats the symptoms of the problem, rather than the underlying syndrome.

Government departments regularly release reports that show their latest understanding of a problem, and lay out the rationale for their current practices. It is less usual, however, for separate departments to be aware of what is happening in other departments, even when they deal with the same user groups. Given this, it can be quite revelatory to combine multiple sources of information at once.

In the 'License to Drive' project (p.62) we saw through analysis of data from the courts that people were being sent to prison for the relatively minor offence of driving without a license. Next, researching the licensing procedure managed by the roads and traffic authority showed us how convoluted and unintuitive the process was. Overlaying this official information with interview data from the experiences of people who had been charged with driving without a license revealed that people were stumbling on unfriendly technicalities and bureaucratic requirements.

Putting data from these different sources together provided a completely new view of the problem and set a different direction for solving it.

Teaming up

People living with a severe and persistent mental illness can sometimes find themselves in trouble. Therapy and medication can help people manage mental illness but sometimes, unexpectedly, and often unnoticed by the person with the illness, an acute episode can lead to a crisis situation.

An acute episode can be incredibly traumatic for the persons themselves and those around them. Family, friends, colleagues and acquaintances aren't trained to help in situations like these, and although they may learn with experience, they are mostly at the mercy of their friend's or loved one's illness.

Police, ambulance paramedics and emergency department nurses are regularly called on to help people who are experiencing a mental health emergency. These professionals each play a different part in the choreography of a crisis response, and together provide a lifesaving service. But a person with a severe mental illness will often actively resist the help that is offered, due to fear or confusion arising from their illness or from previous experience. In these cases, emergency response professionals will struggle to do their jobs, and the person concerned may receive inadequate care – and incur lasting, traumatic memories that create a fear of treatment.

* Based on a text by Mieke van der Bijl-Brouwer

Partners in Recovery (PIR) is an Australian government-funded initiative founded to help people with severe and persistent mental illness by better coordinating services, including emergency and health services. PIR contacted the Design Innovation research centre to design an emergency response that would stop severely mentally ill people from 'falling through the cracks' between services. The Design Innovation research centre worked with PIR on a 6-month program that engaged a range of services that support people when they are in an acute episode, as well as professional caregivers, people with severe mental illness and their friends and family.

Although the initial focus in the workshops and interviews was on the experience of the people with severe and persistent mental illness, one of the fundamental challenges uncovered in the course of this research related to staff. Police, paramedics, nurses and others all accumulate considerable experience, collectively and individually, on how to handle crisis situations and help the mentally ill get access to treatment. Because of the huge diversity, the urgency and enormous complexity of mental health crises, professionals have little choice but to improvise to find the best solution for the situation at hand. But beyond self-evaluation, they have no way of knowing if their improvised response was helpful or effective. Not only does this cause immense frustration for professionals providing crisis care, it limits information sharing and learning – because without

knowing how they 'performed', they cannot be confident in communicating successes and failures with colleagues and the sector as a whole.

Police officer: *"If we do not hear from the person again, there is an assumption that one of three things happened to them: 1) they got better, 2) they moved away, 3) they died."*

We are essentially feeding our efforts into a 'cone of silence' that does not speak back."

The program progressed from problem exploration to solution generation through a series of design research activities and codesign sessions.

A number of solutions are now being trialled, including a multi-agency coaching team. This team collects the experiences of people who have been through a psychotic episode as well as police, ambulance paramedics and all other stakeholders. These experiences will be analysed, and the lessons and success stories will be disseminated back through the organisations.

PREVENT   MONITOR   CALL   RESPONSE   DISCHARGE   LIFE

To protect and serve

When a crime is committed, a number of organisations and entities spring into action. This chain of responsibility, which includes the police, investigators, prosecutors and lawyers, courts and prisons is called the criminal justice system.

The criminal justice system, which has evolved over centuries, is inherently offender-focused. As the name suggests, it exists to bring offenders to justice and its design, therefore, is squarely focused on processing criminals, ostensibly for the benefit of the community. In the last few decades, however, it has become apparent that the criminal justice system's preoccupation with offenders is detrimental to victims of crime, who have described themselves as the 'forgotten people' of the system. At best, the system provides limited solace to people traumatised through crime. At worst, however, it can increase victim suffering by requiring victims to participate in situations that scare and upset them, such as giving evidence in court. This is clearly a problem, since if a criminal justice system cannot help victims then we might well wonder what it is for.

Recent efforts by governments around the world have focused on improving criminal justice systems by 'modernising' them – increasing efficiency in processing cases, cutting red tape, digitising paper-based systems and generally refining and updating systems already in existence. But while these changes might have made things better for the bureaucracy, they have not gone far enough to improve the experience for victims.

So when a team at the Department of Justice wanted to radically transform the justice system in favour of victims, they asked Designing Out Crime to facilitate a human-centred design process that would put victims and their needs at the centre.

Designing Out Crime convened a one-day frame creation workshop (pp.162-179) with a broad group of stakeholders with varied experience of the criminal justice system. Representatives from across government, victims' advocates and police worked together in mixed groups to understand what victims need, and how these needs can be materialised in the system.

Through the workshop process, victims' needs were explored in depth

and translated into aspirational design values, or 'themes'. Among these were 'clarity' (victims should clearly understand what outcomes victims can expect, to reduce stress and trauma), 'safety' (victims should feel protected from danger and free from fear), 'confidence' (victims should know that their situation will be handled by competent people in well-resourced systems), 'compassion' (victims should know that those dealing with their case have their interests at heart) and 'wellbeing' (the criminal justice system should not just process cases, but help victims regain wellbeing in their lives).

From these themes, and a framing and journey-mapping exercise, six solution directions were created and are being developed for implementation.

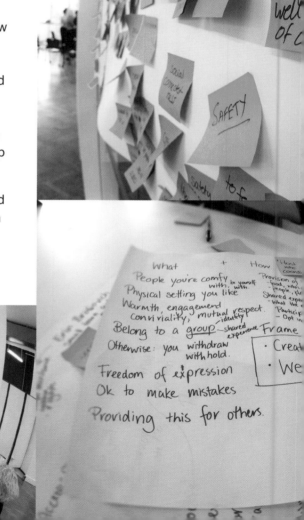

One of these solution directions is a package of initiatives that will help victims to feel safe at court and throughout the court case, feel recognised as part of the court process, and receive practical support to make their experience at court less traumatic.

The solution ideas include timely information about the court process; assistance in attending court (which could include transport, childcare, and support to take leave from work); a concierge service at court to provide directions; a safe physical space at court where victims can wait and have a cup of tea before their case is heard; a designated space in the courtroom; and having their presence acknowledged by the judge.

This suite of solutions, along with the other five (illustrated at right), are aimed at using the criminal justice system to help repair the damage caused to the victims of crime.

Initiatives

1. Improving the court experience.

2. Giving the vicitms greater voice, through expanded restorative justice options.

3. Immediate help for support and recovery.

4. Improved advice on reporting.

5. Targeted help for victims of property crime.

6. Judicial education.

De-separation

Any organisation can be seen as a problem-solving machine, with its structures and processes well honed to solve the problems it normally encounters. The more sophisticated the organisation, the less flexible it tends to be.

These problem-solving machines are based on an established view of the nature and cause of the problems that they set out to solve. The new frames that we create in our projects cut across these assumptions, and inevitably right across the silos in a partner organisation. Thus these projects create new connections within an organisation, or between organisations within a broader network.

It turns out that this is one of the most valued qualities of our approach. Many of our projects bring together people who work in the same workplace or sector, and indeed on the same problem, but who literally have never been in the same room at once.

As we saw in 'To Protect and Serve' (p.82), bringing people together in a room can reveal synergies and connections that empower them to solve the issues they are facing together, in a new way.

This effect is strengthened by the fact that in a reframing situation, everybody is asked to readjust their perceptions, attitudes and – most importantly – their roles. In a reframing process, participants take part as thinking individuals, rather than as representatives of their workplace. This is merely a side effect of reframing, but it is an important one.

Changing lanes

Ashfield is a rapidly evolving, multicultural inner-city suburb well known for its Chinese culture and restaurants. The dynamism and vibrancy of the town centre is played out on a physical stage built predominantly in the 1880s and 1890s, with two-storey shop buildings set cheek by jowl along a 'high street' that was originally designed to accommodate horse-drawn carts and trams. But the high street of yesteryear is today a major highway that bisects the town centre, a swollen river of cars, trucks and buses that overwhelms the friendly, small-scale ambience of the commercial strip.

A network of lanes links the main street with the train station. Every day, many thousands of people use these lanes on their way to and from the trains, shops, library, town hall and the many restaurants. In spite of relatively high pedestrian activity, the lanes convey a strong sense of under-use; disconnected from the vibrancy and hubbub of the main strip, dingy and forgotten, the lanes seem fit for rats, garbage bins and, one might imagine, back street deals. Robberies and assaults have taken place in the laneways.

* Based on a text by Nick Chapman

The local council wanted to make the laneways area less scary and more productive for residents, local businesses, and rail and bus commuters. Council knew what it wanted from the project (vibrant, useable, safe places), but the brief to Designing Out Crime was open-ended, giving the opportunity to get a feel for what would fit this unique community.

At the time of briefing, the lanes appeared to be framed in two possible ways: either as pedestrian thoroughfares or crime hotspots. Neither frame was particularly inspiring. In the course of research, the team noted that although the shopping precincts were consistently busy, shopping at Ashfield tended to be incidental, rather than deliberate; that is, people who used the Ashfield shops did so because they were proximate and convenient, not because they were particularly special or attractive. Armed with this information, a new question emerged: could the lanes become a 'destination' precinct? Focusing on the design values of 'buzz' and 'connection', the team thought about how they could divert some of the

excess energy of the bustling, traffic-choked main strip into the laneways, to create places of interest.

Initiatives include pop-up stalls and night markets; opening the many high-street-facing restaurants at the rear for laneway dining; and using street furniture and pedestrian-friendly paving to create spaces for people to stop and sit away from the overwhelming sounds and smells of the high street. Lighting can help to create a coherent connection between the train station and the services of the town centre: up-lights would be installed in planter boxes and bollards, while festive strings of bulb lanterns overhead would illuminate the major pedestrian routes. Physical realignment of the lanes would reclaim space for tidy-looking, shared garbage disposal areas for residents and businesses, which would in turn give property occupants the opportunity to use their rear yards in more productive (and aesthetically pleasant) ways.

The impact of this project, conducted over a short period, was significant. Council used concepts and images

from the project to stimulate community discussions about renewing the town centre, and to formulate a tender for design of a master plan. Of its participation in the project, council found that the project represented *"a different thought process, a creative process .... which fosters a culture of innovation and creative ideas."*

It would be so easy to just give the laneways a fresh coat of paint, and to upgrade the existing lighting to make the lanes 'better'. This project moved beyond making it better and articulated at a conceptual level what value can be created through rejuvenation.

A place to call home

A few kilometres from the central business district of Australia's seventh largest city, its beaches and vibrant entertainment areas is the Hamilton South public housing estate. With more than 750 dwellings, including units, villas and townhouses, it was established in 1962 as modern, low-cost housing for coal miners and dockworkers and their young families. The architecture is unremarkable among public housing estates built at that time: a predominance of two and three bedroom apartments in three-storey walk-up residential blocks. The land between the unit blocks was designed to be relatively open, undifferentiated and communal, with flat lawns and sparse paving, and shared laundries and drying areas being the only clearly articulated purpose for the space.

By 2011, when a team from Designing Out Crime was introduced to the estate, Hamilton South's public housing residents were identified as being among the most disadvantaged of any community in the country. Rates of domestic violence, alcohol-related violence, burglary and property damage were disproportionately high. Outdoor space was marked by vandalism and the estate's many dark and secluded corners were littered with used syringes, the refuse not of residents so much as opportunistic 'visitors', who used the under protected, semi-public land to deal and inject drugs.

* Based on a text by Rohan Lulham

Around the perimeter, a group of older residents have cordoned off small rectangles of communal space for flowerbeds, and in these little gardens, pride and ownership is blossoming. But the rest of the estate appears oppressive, unloved and alarmingly empty, with few people on the streets in spite of so many people living there. With crime-related problems grist for the mills of local media outlets and political platforms, the government needed to act. Designing Out Crime was engaged to collaborate with a housing design team to develop a program of physical works to address some of the problems. An impending election meant timeframes were tight and funds limited to half a million dollars.

As a decaying housing estate with seemingly intractable social problems, the situation called to mind Oscar Newman's 1970s projects on public housing in the United States that continue to be influential in the broader field of environmental crime prevention. With no definition of public, semi-private and private space on the estate, many problems could be understood in terms of Newman's

principles of access control (stopping people getting at things), natural surveillance (increasing the number of 'eyes on the street'), and territorial reinforcement (promoting social ownership through giving space a function). But the Designing Out Crime team had the sense that a formulaic application of these principles in isolation would achieve little for the residents. It was also evident that the situation at Hamilton South was not an isolated one, and that the project needed not only to solve the problems of this estate but also to demonstrate how similarly troubled areas could be approached in a way that would create nice places for people to live, rather than overly securitised displays of crime prevention principles.

Following extensive research, the Designing Out Crime team designed a suite of physical changes, including new footpaths and gardens to create a coherent and navigable open space, fences to create smaller 'garden rooms' attached to ground floor apartments to meet the need for a sense of ownership, and removal of outmoded laundries that were being used for drug injecting and rubbish dumping. Importantly, the physical upgrade works on site provided an opportunity for new momentum and discussion about what kind of estate the residents wished to live in, and to demonstrate investment in their homes.

Community at the centre

Picture this scene. In a suburb in Sydney's west, children play in a large unfenced park with no play equipment, watched by nervous mothers. Cul-de-sacs and narrow, bending pedestrian lanes surround the park; children could easily skip down them, and be lost. On one edge of the park is a community centre, a grim brown-brick building bearing the scars of graffiti, vandalism and neglect. Inside, the community centre is attractive, and the staff are eager to help, but the building's appearance is so forbidding that the centre is desperately underused by the community.

* Based on a text by Olga Camacho Duarte

Within weeks of opening, the centre had been broken into, vandalised and defaced by graffiti. The government countered this with anti-vandalism measures, including CCTV cameras and a razor wire fence around the perimeter. These measures succeeded to some extent in repelling the vandals, but unfortunately also repelled and alienated the intended clientele – local families – to whom the centre now appeared harsh and uninviting. The community centre looked like a fortified bunker.

Designing Out Crime were briefed to provide solutions that would reduce opportunities for vandalism and anti-social behaviour and encourage the community to use the centre. The team got into contact with the local residents and they wondered how the scheduled activities of the community centre could be better communicated through the look of building and grounds.

'Identity' and 'openness' are the aspirational values that guided this project – values that can only be achieved if the community participates in a hands-on way.

The disused centre was converted into a fun place to play and learn and the community was excited and proud to be involved in implementation. Improvements included a secure play area for children, barbeque equipment and seating for social events, and garden beds for community gardening. A new metal fence was painted in a rainbow of colours – secure but less brutal than its razor-wire predecessor, and a cheerful and colourful addition that could easily be repainted in whatever colour was handy if it were graffitied.

A design was made to remove the bunker-like security bar in front of the community centre's windows, and replace them with a metal artwork. The artwork represents clouds wrapped around the building and is still effective at preventing break-ins, but in a more subtle way.

The residents were glad to be involved in the renovations, and in the following months the community garden started to thrive, as did the programs offered by the centre. The break-ins and vandalism petered out following the community's reclamation of the centre. Years down the track, the project is remembered as a time when the community came together.

Inventive solutions

More often than not, designing is a balancing act between the needs of different stakeholders – and more often than not, coming to a design solution involves some compromise.

But sometimes, designers are able to create a solution that satisfies totally disparate needs, without any apparent compromise. As we saw in 'Community at the Centre' (p.96), conventional security bars on the windows made the community centre look and feel like a bunker. The final design concept for the windows at the community centre was a metal frieze that has the appearance of clouds. While still secure, the frieze has a lightness and an elegance that is friendly and engaging.

In retrospect, such a solution seems inevitably or inescapably good. The struggles of the design process are quickly forgotten in the presence of an elegant design; the solution can feel self-explanatory, beyond discussion – almost natural, rather than the result of toil and invention.

"Of course!", "This is it!" are the common reactions to a design that is beautifully fit for purpose. But while an outcome may appear 'logical' with the benefit of hindsight, there is nothing simple or straightforward about the process of getting there.

It may be that often, such ingenious solutions are just not possible and we will have to settle for a compromise and be ready to defend it. Still, it is crucial to commence a project with the aim of creating a solution that helps to realise value for all; starting at a compromise means that compromise is the best we can achieve. We like to think that some of the framings presented in this book have achieved a high level of coherence and integration. Inventive solutions are worth striving for with all our might.

Houses for people

Public housing has long provided shelter for people in need. Once, whole suburbs of public housing were built to provide accommodation for low-income, working families, who often occupied the houses like homeowners – doing their own repairs and renovations, and passing the leases on to their children and even grandchildren.

For much of the last century, public housing served this group relatively adequately. But today, the bungalows, townhouses and walk-up flats and the suburbs they fill are home to people who are not only poor, but have more complex needs and are reliant on other government services – young single parents, the elderly, people with mental illness, people with a disability and people recently released from prison. Neither the dwellings nor the suburbs were built with these users in mind, and they are no longer fit for purpose. As a result, public housing has gradually become a place of grudging last resort for poor people, a place to live when every other option is unobtainable.

* Based on a text by Olga Camacho Duarte

Suburbs that are comprised predominantly of public housing today attract a stigma that further alienates residents. This poor reputation is based on higher than average crime rates – the result of very complex factors – and a measure of prejudice and ignorance that seeks to oversimplify the causes of crime. Disproportionately negative media representation perpetuates the stereotype and feeds a vicious circle.

The government's public housing department engaged Designing Out Crime, the local council and a social work school at the University of Western Sydney to form a three-year partnership to see if they could make public housing work better. The partnership's stated intention was to focus on the physical environment and the amenities of the area. Social work students would research issues and feed their findings into the design processes.

Early in the partnership a four-day 'deep-dive' workshop was held, with 30 participants from a diverse range of organisations. After four thought-provoking days, four themes were identified that would help inform the partnership's projects: identity, aspiration, empowerment, and reflection.

The following projects are among many that were done as part of the partnership.

Local Shops

A small row of dilapidated suburban shops was struggling to function. A fire had gutted one of the shops and others had been vandalised and robbed. Patronage was low, yet there were few alternatives for locals.

The team explored how the shops could be converted from a problem area into a place that the community could be proud of. Solutions included building residential apartments on top of the existing shops to create a place that would be busy and active day and night, holding markets in the underused car park, and using vacant shops to create a community hub where services could be conveniently co-located.

All of the stakeholders agree that the approaches recommended would be effective, however current the rules governing building height do not allow the redevelopment. Any future work on the shops needs to engage the planning authorities, as changes to the rules are needed in order to implement the solutions.

Pedestrian tunnels

These suburbs were designed for people with cars, which many of the current inhabitants do not have. The distances between essential services and basic conveniences – shops, bus stations, schools, hospitals – are considerable. The public footpaths, where they exist at all, are not continuous or meander indirectly from one point to another, as if the people who live there have all the time in the world to wander to the shops. Swathes of grassed areas with no particular purpose complete the impression of vacancy and absence that defines public space in this area.

Under the quiet roads, pedestrian tunnels as dark as caves link one open, empty space with another. Locals avoid this unappealing and carelessly designed infrastructure because it looks dangerous, and because they cannot tell where each pedestrian tunnel leads. As is the fate of disused space, the pedestrian tunnels now represent a legitimate danger. They are vandalised and littered with broken glass and syringes.

There has been a proposal to close the pedestrian tunnels altogether, but the housing authority are looking for other solutions.

The team looked to the community for answers. If the suburbs were built for a different time and a different population, who were the new occupants? What did they need and value in a public space? A demographic profile and a series of interviews identified that the new residents were, predominantly, immigrants from a few very small Pacific nations. In talking with them it was clear that community and family gatherings are very important, and that at these get-togethers the preparation of food is an important communal activity. The houses they live in don't cater for these large family gatherings.

Through their research, the team found a match in the community's need for outdoor space, and the empty spaces either side of the pedestrian tunnels; and since the pedestrian tunnels cannot be altered or improved without considerable cost, the proposals focus on the land on either side of them.

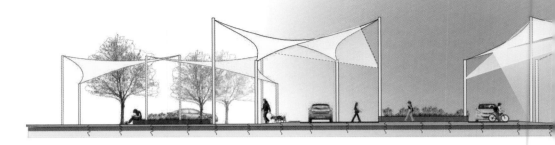

The team designed recreational areas that included communal cooking facilities, shade, shelter and play equipment to support gatherings of friends and family. By adapting public places to the needs of residents, the vast, empty open spaces in the suburbs can become used and lived in, and the pedestrian tunnels can, at last, serve a purpose.

Town centre

A common issue in the sprawling, satellite suburbs is lack of population and lack of activity in public spaces. However, this is rarely a problem in the town centres where people work,

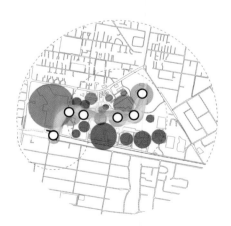

shop, visit doctors, go to the library or on cultural outings, and meet friends. Streets are busy and vibrant and each square metre of space has a designated purpose.

Recognising these positives the council – which was a great advocate of taking a strengths-based approach of the partnership – worked to strengthen the town centre in hope of lifting the rest of the suburb. Council had two development processes underway and plans were being made to upgrade the pedestrian mall in the town centre, and the nearby sports park. The partnership turned its attention to these developments.

While the town centre is busy and active, crime maps showed that it is still a hotspot for violence, property damage and robberies. Litter was a

particular concern. Past attempts to solve rubbish problems had included adding more and more bins, since it was assumed that people littered for lack of them, but this has not worked.

Apparently logical yet failed remedies for problems in the public domain were by now familiar territory for the partnership. It was clear that we needed to identify the problems behind the problems. Answers were found not in the town centre, but by looking further afield.

The sports park is only a short walk from the mall. At the park, there are playing fields, some rare native forest, a very well-utilised indoor sports stadium, and a pedestrian pathway that leads to a hospital. However, the park is obscured by buildings – many of which have their 'backs' to it – and many residents and visitors simply don't know it is there. Sports clubs scour the ground for used syringes before play, as there are great stretches of time when the playing fields aren't used for sports. People have said they fear the park at night because it feels so isolated.

The team could see a great many options that could improve both the park and the mall. A large number stakeholders, with similar aims and objectives could be involved in a solution. All that was lacking was a coherent concept to bring them together.

The team designed a pedestrian connection concept, which would tie together the route from the train station, passing through the pedestrian mall of the town centre, to the

sports field, ending at the hospital. Connecting the services that are already there creates a whole that is greater than the sum of its parts. The sports ground would be augmented to cater for a broader array of sports, and personal training infrastructure installed to create a place that would be used all day round.

The mall and the sports park are both now being improved by council. The projects have informed the development process and have helped council reconceptualise not just the physical space, but also how the spaces could be managed to create the best possible experience for the people who use them. The ideas developed for different project sites are being used for the partners' social, cultural and urban plans.

The projects have helped facilitate the collaboration between the partner organisations externally, and the various divisions of the partner organisations internally. The new frames are a catalyst for integration.

Barangaroo

Nature & the city

Barangaroo is back in business. The former industrial wharf on Sydney harbour had drifted into distant memory until a design competition in 2006 sought to repurpose and redevelop the site into a high-profile landmark for the city. The competition process produced a concept design that consisted of three main areas, one of which is a park that recreates the feel of the original headland during Aboriginal settlement. This sizeable park is known as Barangaroo Reserve.

Designing Out Crime were engaged by the developers of the site to advise on the crime prevention aspects of the concept design. This approach is highly unusual – crime prevention is typically only considered in the late stages of the urban planning process. Given the unique opportunity to shape the park's design early on, the team were faced with two questions:

> What does a safe urban park feel like (ideally)?
> How can design actually achieve this?

* Based on a text by Kim Wan

These questions required a deeper understanding of the social and design contexts and the interaction between these realms.

The project began with a study of six well-established urban parks around Sydney. The team interviewed park managers on their experiences with day-to-day and seasonal park management, park users and their habits, the evolution of their park's design, and any issues with crime.

From the interviews it became apparent that urban parks were in very high demand as outdoor recreation spaces. Issues with crime were mostly linked to perceptions of safety; actual incidents were isolated and often stemmed from activity related to nearby cafés and sporting grounds. The increased popularity of open-air exercise classes and cycling, and a growing number of festivals had increased use of the parks, and occasionally these activities conflicted

with the more traditional, passive park activities – picnicking, lying on the grass, contemplating nature. It emerged that all of the parks had a very different character and feel at different times of day, with distinct busy and quiet times.

Once the detailed design of Barangaroo Reserve had taken form, the designers used this research to underpin a discussion session with a very broad group of stakeholders with experience in park design and management and crime prevention, including police, council, parks and gardens authorities and public space managers.

Stakeholders were familiarised with the park design before forming mixed groups, where they were asked to share personal experiences of parks as a way of empathising with public patrons. By taking off their organisational hats – even for a moment – participants were awakened to the value of human, experiential

knowledge rather than purely the professional expertise on which they are normally encouraged to rely. The qualitative data that emerged was rich and valuable for the final stage of the workshop where participants discussed and critiqued the designs together, using a combination of newly-revealed 'human' knowledge and professional experience.

For the design discussion, each group created scenarios of likely future use and considered how the park would fare under these conditions.

The scenarios revealed concerns about things like site capacity for events, managing movement on a site with permeable borders, and problems with sightlines in dense forest areas. This discussion was then distilled into design recommendations for Barangaroo Reserve that addressed crime prevention concerns but also covered park management, hours of operation, gating and activities and – most importantly – how to make different groups of visitors feel welcome and at ease.

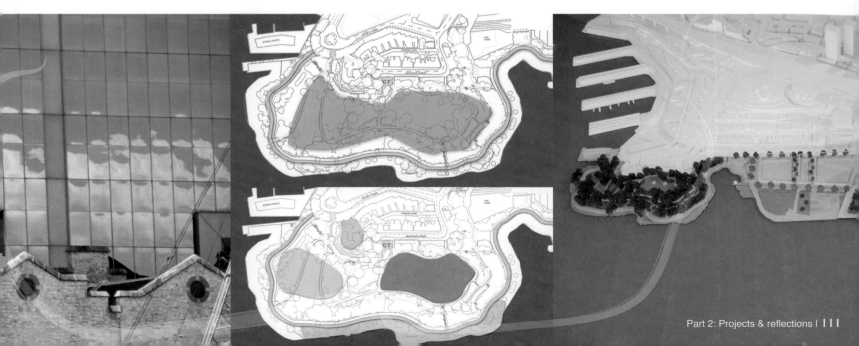

quiet
1 on 1   Collaboration.   Peer.

PIN   PIN

IWB.

T2   T

T2

Intensive Learning Centre.
Learning Module.   → Educator positioning
2 WINGS. + multiple learning/teaching modes.
one person to every one (IPE)

Thinking inside the box

Prisons make prisoners. It's alarming, but true: the single biggest risk factor for ending up in jail is having been to jail in the first place. Rates of recidivism (reoffending) among prisoners are estimated at around 80%; that is, as many as four in five prisoners can be expected to reoffend upon release.

Another significant predictor of criminal behaviour and subsequent incarceration is a lack of educational and employment opportunities. The overwhelming majority of assessed (male) inmates in Australian prisons have reading, writing and numeracy skills well below expected adult levels; put differently, educational attainment among the prison population falls drastically short of the minimum standard required for opportunity, self-determination and meaningful participation in society.

These statistics alone predict a fairly bleak trajectory for prisoners, not to mention the broader society that suffers when further crimes are committed. Preventing offending in the first instance and reoffending in the second has therefore been a priority for government in recent years.

* Based on a text by Rohan Lulham.

In 2012 the government wrote an inspired brief to build an 'Intensive Learning Centre' in a maximum-security prison. The brief envisioned a place where educators and up to 40 inmate learners would engage in project-based learning, using twenty-first century technology within a supportive, therapeutic environment. The intention was to provide high-risk inmates with an opportunity to gain new knowledge and skills and greater confidence in their abilities, making them more employable upon release as well as better prepared to complete prison programs that treat the causes of their offending. In this way, the government aims to break the cycle of recidivism.

Typical prison architecture and design seeks to separate, contain and restrict. The government knew this to be incompatible with effective learning, but did not have the internal expertise to design an environment that would transcend the prison walls and promote interaction, transformation and aspiration. Designing Out Crime was approached for the job.

The research process, which was immediately recognised by participants as unusually (and pleasantly) inclusive, included talking extensively with prison staff, management and inmates to understand the needs and harvest the experience, skills and knowledge of each. The resulting program space and philosophy reflected this diversity of input and the freedom inherent to this process.

Interpretive and visioning devices such as themes and metaphors were widely used in the project, both to design and to communicate ideas. A defining value was 'connection', which referenced the aspiration for inmates to (re)connect with family,

referenced cloud technology and the idea of universal and ubiquitous knowledge that is not place-bound.

The cloud metaphor was expressed in the design of the classroom roof, which is curved and slopes gently upwards at one end, to give learners a view to the clouds. The metaphor and its manifestation in the physical design was particularly appreciated by the inmates, one of whom subsequently declared that he wanted to become an architect.

The final designs were constructed by inmates and hoisted over the fence at the maximum-security facility. The first cohort of inmate learners graduated in December 2014, following an official opening by the Attorney General. It was a celebration of achievement and the promise of opportunity, aspiration and transformation that is rarely seen within a prison environment. In this learning environment, the inmates are given the chance to imagine the new and different life they might lead when they are back in society.

friends, workplaces and society more broadly through education, work experience and personal growth. A powerful metaphor for connection was the 'cloud' metaphor, which referred to the way in which clouds traverse the territory of the sky, and how the same cloud that is seen by inmates can, moments or minutes later, be seen by people 'on the outside'. This metaphor also

Seduction

All design works by seduction: it needs to persuade or seduce people to 'play along' and use a design as intended. Deep and intimate knowledge about the user group is completely crucial to this and comes from proactively seeking to understand people's behaviours, needs and values.

In a project exploring fear of crime and disconnection in a high-rise housing estate, some visiting students from Korea University took us at our word when we challenged them to embed themselves in the situation. They accepted the invitation of an elderly Korean resident (whose name translates to 'Harmony') to live with her for the duration of the project, and the insights they gained brought immense value and understanding of the needs of residents to them and the other student designers.

The students followed Harmony around each day, observing her routine and interviewing her about her life. They noted the details of when she woke up, how and when she cooked her meals, when she went out and what she did while she was out.

They observed her interactions with neighbours and gained deep insight into Harmony's physical, social and emotional needs. Their findings were surprising. Harmony lives a contented life, but she has much to share and nobody to share it with.

The value of design lies in the creation of a 'something' for 'somebody'. Creating this connection means shaping high-quality relationships. It became clear to the student designers that while there was social connection between cultural groups within the estate, there was very little connection between the estate and the surrounding suburb. The solutions put forward including converting the grounds of the estate from lawns into cultural gardens; each garden would celebrate one of the many cultures represented among the residents in the estate. Then, neighbours from the surrounding area would be invited to visit the gardens, to be seduced by their physical beauty, with the ultimate aim of making new connections with the residents.

Opening the door

Were it a matter of simple addition, an organisation made up of compassionate, intelligent, cultured and wise employees would be a formidable force for good – the sum of this individual excellence. But in his seminal work *Images of Organization*, Gareth Morgan describes organisations as prisons with imprisoning behaviour. And to a certain extent, this is true: since organisations structure the way we work together, they stimulate certain behaviours and repress others.

In this way, organisations can encourage or force us to 'lock away' or 'lock out' certain aspects of our personalities, attributes, interests and concerns so effectively that it can sometimes seem that we leave them at the door when we enter the office each morning. This can be good, or it can be very, very bad, as illustrated in Zimbardo's famous and disturbing Stanford Prison Experiment.

We have heard from our partners that a frame creation process can help release both great frustrations and great qualities in people – qualities hitherto unexpressed or simply not valued in the day-to-day running of the organisation. We have been amazed, many times over, to discover the energy that is latent among a group of employees who have been given the freedom to come to work with all of their talent and personal excellence blazing.

There is so much untapped potential in organisations, and being in the position to tap it (as we often are in our projects) is humbling, and a great privilege. Frame creation is an open process that allows people to flourish, that opens the door to a safe place for new ideas.

Morgan, G. (1986). *Images of organization*. Beverly Hills, Sage Publications.

Zimbardo, P.G. (2007). *The Lucifer Effect: Understanding How Good People Turn Evil*. New York: Random House.

Look, and look again

In this book we look at design through the lens of problem solving. But that raises the question: what is a problem? We perceive a situation as problematic when we strive for something and continually fail to attain it. Something is holding us back, or forcing us back, and we feel stuck.

So a problem is man-made; it is created and felt by people. This inherent subjectivity can be very problematic in itself. Sometimes we might perceive a problem when there is little ground to do so – where in the real world, there is no problem – or we might find out (to our cost) that the problem is different to what we initially thought. That might lead to misguided problem solving efforts; or, if we are unlucky, or aren't thoughtful, will actually make things worse.

Media can easily create these misperceptions by focusing on incidents that make for juicy stories when real news is in short supply. And that is dangerous, because if a problem is misframed, the wrong perception can easily be set and it can be hard to change.

In the 'Houses for People' project (p.102), we saw how a workshop held at the start of the partnership helped stakeholders discard the notion of 'troubled' and crime-ridden suburbs. For the next three years, the projects done in this partnership were guided by aspirations rather than deficits.

Integrated living

Traditionally, the government policy in Holland, as in many other countries, was to house people with an intellectual disability outside society. They were cared for in institutions that were often beautifully positioned in wooded, secluded areas of the country. While society took pride in the quality of the care that these institutions provided, the people in them were also hidden from the general public.

Recently, this policy has been reversed: the contemporary wisdom is to encourage intellectually disabled persons to live their life as 'normally' as possible. The translation of policy into practice has included rehousing the intellectually disabled to live independently in towns and cities, with some support from a network of caregivers. This new ideology has had huge and often disastrous consequences for the intellectually disabled. Although their physical isolation ended when they entered the world of 'normal' people, mental isolation has persisted; the bodily relocation from institution to private apartment does not assure inclusion in society. They have trouble integrating into their neighbourhoods and struggle with the pace and character of city life. Their neighbours generally ignore them, lacking the time and patience in the frantic rhythm of their own busy lives to take the extra trouble to interact. This leaves people with an intellectual disability stranded in their apartments, desperately lonely.

The government commissioned Peik Suyling's Young Designers Foundation to undertake the 'Integrated living' project because the issues that are facing these people are complex and require creative solutions that potentially involve many stakeholders, spread throughout society. The initial partners included a disability services organisation, a medical infrastructure institute, a major project developer, a building corporation, and a new media think-tank, working alongside fourteen young artists and designers.

In the course of early discussions, the initial question posed by the ministry was drastically redefined. The government had, by default, cast the problem in terms of the need to 'care for' the mentally disabled, while the artists and designers immediately approached the situation in terms of their abilities. This was a breakthrough, because thinking about abilities opens up the problem arena to considerations of how people can and do contribute to society (the following quotes are taken from Suyling et al [2005]).

"The designers were right not to accept the fact that intellectually disabled people live outside society. They understand that the intellectually disabled have their own ambitions." (Ministry of Health)

The question to be answered then transforms from one about 'caring' for people, into the challenge of finding ways in which the contributions of people with an intellectual disability can be shaped and facilitated.

"Some intellectually disabled people are at home a lot, so they can make a positive contribution to the social surveillance in the community. The safety and security in the community also increases through the presence of nurses and carers..."* (partner organisation)

Some of the young designers engaged deeply and personally in the life of the disabled people to deepen their empathy and get a feel for where solutions might lie.

"(in my research) I faced some problems, since this group of mildly disabled people has difficulties in verbal expression, and social interaction. They are often illiterate, so you cannot send them a questionnaire or have a conversation in the way you are used to. Even communication by telephone led to strange misunderstandings." (designer)

In a thoughtful essay, one project participant remarked that her own experience in interacting with disabled people was rooted in the experience of growing up in a small village. Her year group at the village school was very diverse; some children were mildly disabled, and she and the other children naturally learnt how to relate to them. Since moving to the big city – ostensibly a much more diverse population – her social circle became more limited and she did not meet such people anymore.

Other designers explored the roles of the institutions and caregivers. They experienced first-hand how hard it is for a willing and committed person even to get access to people with a disability. In the course of the investigation, it became clear that the over-protective attitude of the care institutions and the care workers contributed significantly to the isolation of disabled people. Inadvertently, the responsibility to care had been extended to protecting people from their new environment, including issuing warning signs not to open the door to strangers. Care institutions and care workers hadn't come to grips with the fact that in this new living situation, their 'patients' are no longer completely protected, and their lives cannot be controlled. Navigating the dangers of normal city life is difficult and uncomfortable, perhaps especially for a medical organisation where risk is virtually intolerable and is managed as such. This insight alone uncovered many new possibilities for improving the integration of people living with a disability.

Often, the issues gained depth and humanity (away from mechanistic or technocratic lingo) by being rephrased as personal questions to the project participants:

'Are you, as a non-disabled person, integrated into your neighborhood?' (partner)

'The real questions relating to this project are: why do people want to meet each other? When do they become friends?' (designer)

At the conclusion of the project there were some promising perspectives.

Not 'big solutions' to 'the big problem', but rather departure points that together provide a fascinating map of possibilities. There are many issues and avenues that need further thought and discussion – the role of 'care' and the way care is institutionalised in our modern society surfaced as a major theme. And the experience of being involved in this unconventional problem-solving exercise has had a profound impact on the project partners.

Suyling P., Krabbendam, D., Dorst, K., (eds) (2005) *More than 8 design ideas for the integrated living of mentally handicapped people in society*, Ministry of Health Wellbeing and Sports, The Hague.

Radical frames

Amsterdam West is an urban district where problems have accumulated, many say. Less attention is given to the fact that there also is an accumulation of potential. Amsterdam West as a municipal administrative entity has developed an approach towards public management that is participatory and design-oriented. It brings together different teams of citizens that – quite literally – take to the streets to explore, define problems and design made-to-measure approaches to urban issues.

One of the more apparent problems in Amsterdam West is the way in which young men of Moroccan descent manifest themselves. Where in previous years, involvement in crime was seen to be an attractive way to 'be someone', increasingly, jihad has come to fulfill this need as well. Real concerns about the plight of Palestinians and others throughout the Islamic world have fuelled a radical opposition to anything that could be interpreted as a threat to Muslims.

* Based on a text by Rob Ruts and Dick Rijken

It was in this context that the municipal government of Amsterdam West commissioned the Hague University of Applied Sciences to look into ways in which the community could be better equipped to deal with radicalisation. The university's department of safety and security management offered international students a laboratory-style project to explore new ways of approaching radicalisation. During ten consecutive weeks, the students framed and re-framed the notion of radicalisation, gradually dismantling it through a series questions such as: is radicalisation a uniformly identifiable process that characterises a specific group? Is it even a process at all? Or is it a label that identifies the outcome of a complex networks of circumstances? An interesting thematic analysis resulted. Starting from the concept of 'radicalisation', students identified the underlying human themes 'hope', 'anger', 'frustration' and 'vulnerability'.

They then analysed and explored the meaning of these themes deeply, using a set of questions and methods developed by their professor, Dick Rijken.

To understand the nature of a theme, they first asked generic questions like: what other concepts is this theme related to? (e.g. what is the relationship between 'hope' and 'anger'?) What personal motivations can be involved? To make the discussion more specific, the students then reconnected the theme in the problem context: how does a theme (like 'hope') play out for the Moroccan boys and girls in Amsterdam West?

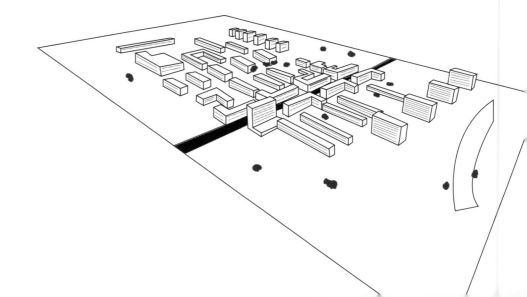

To understand the structure of a theme, they used a comprehensive set of questions that span the dimensions of a person's psychology:

- The spiritual dimension - does the theme relate to a person's deeply felt beliefs or values?
- The emotional dimension - what emotions are involved? How?
- The cognitive dimension – what mental models, knowledge, learning processes, thinking, &c is this based on?
- The physical dimension - does the theme have physical/biological aspects?
- The social dimension - how can social relationships affect the theme?

Then they considered the dynamics of a theme, how the personal experience of a theme can change over time. Websites like www. emotionalcompetency.com (which has flowcharts for themes like 'fear') help to map this out, and give an idea of the factors that have a negative or positive influence on this development.

To attain a deep understanding of a theme it needs to be researched thoroughly. Here the student groups split up and researched the theme from many different angles. Scientific literature can be useful, but also hard to use as it tends to be very abstract and specialised. Philosophy can be a great starting point for understanding themes because some philosophers devote most of their life to thinking about a specific theme (Marx on 'alienation', Levinas on 'otherness' and 'responsibility', &c.)

The Arts – music, visual art, theatre, film – are excellent sources of theme knowledge because they provide universal insight while being very concrete at the same time. Scenes from films can sometimes very efficiently evoke the emotions that are connected to specific themes.

And then, of course, the students' personal experience is a key source of knowledge. Examining personal experiences of the theme or interviewing people about theirs (asking questions such as: have you ever experienced 'anger'? What triggered it? What did it feel like? What were you thinking? What changed it? What did you do when you felt anger?) helps to integrate the knowledge gained from all the other sources.

After this in-depth theme analysis, the students connected their four themes in a scenario that helped to give the themes meaning, as in: 'when young people find themselves in a society where they cannot realise their hopes and dreams, and where they feel discriminated against, their dreams decay into frustration and anger, at which point they become vulnerable to ideologies that reject the values of this society and offer alternative ambitions'.

The students used the deep theme analysis as preparation for fieldwork. In Amsterdam West, they were introduced to an urban environment that was new to most of them, and they heard the stories of people working there. One of the dominant conclusions was that radicalisation can be thought of as a reversible set of processes that is fuelled by needs that young people have, independent from their ethnic, social, cultural, religious backgrounds. Young men and women – all of them – are in search of identity, kinship, ownership and ways to make a difference. As one expert on radicalisation said: *"It is a matter of demand and supply. When the environment in which these kids grow up has nothing else to offer than the jihad as identity-giver, the temptation will be enormous to take that as the context for the search for identity. What is needed is a less destructive subculture that provides an platform for criticism and rejection of society, and where young people learn to deal with frustration and anger."*

It was clear that the interventions should focus on the 'supply' side, creating ways for young people to associate themselves with the city and use it to fulfill their needs for identity, kinship and ownership. At the end of the project, the students presented a suite of interventions in the form of sports and martial arts, but also cooking and dance that would involve different generations from the community. Solution ideas also included opportunities for experiences such as competing, winning and losing, dealing with defeat, and challenging oneself and others in safe ways. These were explicitly not presented as 'anti-radicalisation' measures, though they could be measured as such.

Layers of reality

To successfully reframe a problematic situation we need to be able to scratch the surface. A superficial view of a problem situation will inevitably put us on the path of solutions that have been tried before – unsuccessfully, or only partially successfully (otherwise the problem would not still be there for us to solve). Seeing the layers underneath requires a certain distance, and the facility to look carefully and deeply at the problem situation without falling into the ways of thinking of any of the problem's stakeholders.

In their day-to-day practice, organisations most commonly see 'practical' problems – those that can be solved by the practical means available to them. But when the same problems are encountered time and time again, there is sure to be a hidden problem layer. Then we should ask: what have we here?

We use a subtle process of theme analysis that is informed by the practices used in hermeneutic phenomenology (van Manen, 1990) to help us locate these layers. In hermeneutic phenomenology, theme analysis is used to form a deep understanding of a written text. We use it in our work to find and 'read' the layers in real-world situations. As we have seen in the 'Integrated Living' project (p.122), the isolation and heartbreaking loneliness of intellectually disabled people cannot be addressed without a deep understanding of how notions of 'care' and 'control' are linked in the DNA of the care institutions.

In our projects, we work through this process together with problem owners and stakeholders. We then return to the 'practical' problems with this deeper understanding, and together explore how day-to-day practice can be changed to solve the problems.

Van Manen, M. (1990). *Researching lived experience: Human science for an action sensitive pedagogy*. Suny Press

Aloneness

A project that was sparked by the crushing loneliness of people with an intellectual disability living in an urban environment (see 'Integrated Living', p.122) has led to an interesting sequel, years later. In this project the theme of loneliness itself took centre stage, as one of the contemporary problems that leads to unhappiness and suffering by many, and one that governments, health authorities and non-government organisations are grappling with.

The starting point for this project was that the single word 'loneliness' actually hides many meanings. And the word is problematic, since it is tendentious, associated with shame and stigma: you are a loser when you are lonely, which makes the situation all the more painful.

The young artists and designers working on this problem confronted it head-on by going out on the streets with sandwich boards painted with personal statements about their experiences of loneliness. This disarming approach led to many good conversations with passers-by, and a small documentary was made that showed how perfect strangers, interviewed on the street on a rainy day were very thoughtful and open in their understanding of the issue. They saw loneliness as a normal part of human existence, and even a situation to be embraced and valued – one woman explained how the utter loneliness of a walk in nature

had led to a change in her perception and an unexpected feeling of blissful connectedness. The idea that loneliness can be a portal to reflection and insight came up several times. Apparently there is more to loneliness than meets the eye.

This clearly is a theme worth understanding more deeply. Together with the Verwey-Jonker Institute and the Stichting DOEN, the Young Designers foundation set out to explore this theme through artistic and designerly interventions. These interventions took place over a period of several years, and took many forms.

In one example, an interaction designer created workshops in which people were asked to explicate the different types of loneliness they had experienced in their life. As a prompt, the designer created a typology of loneliness using the 22 Inuit adjectives for 'snow'. By replacing the word 'snow' with the word 'loneliness' in these phrases you get 'gritty loneliness', 'drifting loneliness', 'melting loneliness', 'light loneliness that is firm enough to walk on', &c. – beautifully rich and poetic sketches of loneliness that are nearly or possibly recognisable in common experience. These stimulated a subtle discussion on loneliness in the workshops.

In another intervention, an artist created an intricate questionnaire in which interviewees are led through a series of questions on loneliness that both help them define their own experience in a sophisticated manner, and wonder at the broad array of possible lonelinesses.

In a third intervention, a photographer focused on her lived experience of loneliness. Her intervention, 'The beauty of loneliness', consisted of a moving series of photos captured at moments of utter aloneness, to express the special feelings of heightened perception and sensitivity that such moments can produce.

After the theme of loneliness was thus explored, it gradually lost its negative connotation and became more aligned to the cherished Eastern spiritual concept of 'harmonious aloneness'. Perhaps one should not always try to 'fix' loneliness by eradicating the opportunities to feel lonely, but also appreciate that loneliness is a natural part of life. This radically reframes the way we might approach loneliness in society.

methods

3.

Introduction

By now it will be clear to the reader that there is no single formula for designing for the common good. Not only do these projects differ in scale, size and subject, but each is heavily influenced by the content of the project and its specific context. This requires great flexibility from the practitioner and an ability and willingness to improvise where necessary.

Nonetheless, it is important not to be so flexible as to lose all structure. We have found it useful to plan projects in small building blocks that can be arranged and rearranged as necessary for the problem at hand. These building blocks are our 'method cards', a set of loose cards that were developed and designed by Designing Out Crime staff. The methods have come either from the individual discipline areas represented among the staff (including architecture, psychology, urban planning, history, industrial design, criminology), or were developed 'on the fly' to suit a particular project.

Each card includes a brief description of the particular method or technique, along with a 'how-what-when-why' guide to its application and an example of a past project where it has been used. Also on each card is the name of the method 'owner', an experienced team member who can be contacted for tips and tricks on its use.

The creation of these cards was a worthwhile and cathartic group exercise, providing us the rare luxury of analysing past projects and identifying precisely and exactly the tools and techniques we have used – and the merits and pitfalls of each. Partner organisations often ask if we can share our method cards. This is less beneficial than it might sound, however, because the process of mapping the methods together in a team is nearly as important as the cards that result.

But since this is a book in which we share our work through our projects, it seems fitting also to share the methods we have developed.

So here they are.

Ask nicely

WHY: Want to know what somebody thinks about something? Sometimes it really is as simple as asking a question and listening to the response.

"I like to listen. I have learned a great deal from listening carefully. Most people never listen."
– Ernest Hemingway, *Across the River and Into the Trees*

HOW: There are many ways to interview. Consider the following approaches:

Semi-structured – This more casual approach to interviewing provides the opportunity for open, two-way dialogue. The interviewer explains what they are interested in discussing, and the conversation proceeds from there. The interview is guided by a pre-prepared framework.

Personal story – Ask your subject to tell their story. What life events have brought them to where they are? There are many 'ways in' to this approach and you can be creative and thoughtful in soliciting a story.

Regardless of the interview technique you choose, it is wise to develop a plan for the types of questions you want answers to, and to take notes in a way that doesn't hamper the flow of conversation.

Having recorded the interviews, the next step is to reflect on what the interviewees were *really* saying. There are a few methods for this (see the method 'Understanding Perspectives', p.148) for more detail on how to use interview data to extract themes and create deeper understanding.

it was a clear black night

a clear white moon

Write the story

WHY: Writing a narrative helps both writer and reader develop empathy and a good conceptual understanding of a problem as it presents, or how a new solution could play out.

"I never travel without my diary. One should always have something sensational to read in the train."
— Oscar Wilde, *The Importance of Being Earnest*

HOW: There are many narrative forms. A couple of versatile forms that we have used in projects are:

Short story - Construct a short story that is played by actors in the problem context. The story can be drawn from personal experience, or it could be a fiction starring people you know, or a totally invented scenario. Express the experiences and feelings of each character, relationships between them, actions and consequences. Start your story at the 'beginning' (with the genesis of a problem), in the middle (as a problem scenario is unfolding), or at the 'end' (with a possible future solution).

News article or press release – Project into the future and write a press release or news article announcing a new and wonderful solution to your problem. Be wildly inventive and unrealistically positive in describing what your idea looks like. Explain how you (and others) got there, invent quotes from important stakeholders, and generally go to town.

Get a feel

WHY: How we feel in a particular place at a precise moment in time informs our thoughts and future choices, and puts us in a better position to empathise with how others may experience that situation. This tool is particularly useful for place-based problems.

"The world is full of obvious things which nobody by any chance ever observes."
– Arthur Conan Doyle, *Sherlock Holmes*

HOW: Choose a 'trouble' spot – such as the location of your problem, if it is place-based – or somewhere that you really love, and explore how each place makes you feel. Using this method in more than one place and time (day and night, for example) can provide a useful comparison. Bring a companion for safety if you need.

Go a bit Zen. Close your eyes, take a few deep breaths and concentrate on what your senses are telling you. Open your eyes and have a wander around, noting your sensory experience and emotional response to the place. Observe how other people appear to be feeling. You can try role-playing this exercise – putting yourself in someone else's shoes and imagining how they might feel. Record your experience in detail.

Point and shoot

WHY: Photography is an invaluable, essential and ubiquitous tool in today's highly visual world. Photographs can be used for research, to document a problem or scenario; or as a communication medium, to help explain a concept or sentiment, or persuade others.

"A picture is worth a thousand words"
– Chinese proverb

HOW: Since photographs are so versatile it is wise to take many, and often, during the course of a project. In a place-based project, photographs may constitute the primary record from a site visit (see method 'Where? Here?', p.154) and a starting point for research. In any project, photographs (of people, places, scenarios) are a powerfully emotive way to present a case for change or to convey an impression of an ideal future solution.

The method or mode in which photographs are presented is important.

A photo-essay is a visual narrative, a curated series of photographs that tells a story about people, places, events, relationships and interactions. Slideshows are a convenient way of telling your story visually, and may include captions or narrative script alongside images, or not. A 'pecha-kucha' slideshow is a popular format which consists of 20 slides, strictly timed at 20 seconds per slide. Alternatively, a feature photograph or hero shot can be all you need to make your point. Find one image that says everything.

Understanding perspectives

WHY: If a problem is stuck, chances are that the stakeholders each have their own firmly held agenda and that these agendas are in conflict. Mapping the different perspectives of stakeholders can help uncover these agendas.

"It is a narrow mind which cannot look at a subject from various points of view."
— George Eliot, *Middlemarch*

HOW: Use methods such as 'Ask Nicely' (p.140) and 'What Do You Know?' (p.156) to find out what stakeholders think and care about.

Write the problem in the centre of a page and map out each stakeholder, their relationship to the problem and anything else you know about their attitudes and perspectives. Discuss.

"To perceive the world differently, we must be willing to change our belief system, let the past slip away, expand our sense of now, and dissolve the fear in our minds."
— William James, *Pragmatism*

Painting by numbers

WHY: Quantitative data (data that can be measured in numbers) can provide a broad understanding of an area of enquiry. Governments are generally very good at collecting and sharing data on the things that they do. Gathering statistics about a topic is a useful early exercise and, combined with other research methods, can help to build a solid understanding of a problem.

For example, data on national demographics such as age, gender, employment, health and educational attainment can be compared to local data to help build a picture of local strengths and challenges.

HOW: Find reports and publications from people and organisations who have gone to the trouble of gathering and publishing data on topics that intersect with the problem. There are plenty of sources of statistics and reports. Some organisations specialise in pulling together this data, while others publish summary statistics on their own area of interest.

Collate the quantitative data you have gathered (along with, ideally, qualitative data) and share it with stakeholders to start new conversations about the problem.

Stats only the beginning

WHY: Many problems come to us in the language of criminology, one of many disciplines that underpin our work. Criminologists often begin investigating a problem in terms of the types of crimes that occur, the locations in which they occur, the times of day and day of week, and the victim and offender characteristics. From this data they identify patterns and trends to guide further research.

HOW: If the starting point for your project is a specific location, look at 12 months of geo-located data of the location and surrounds to see where crimes of different kinds are more concentrated.

If the problem is not situated in a place, exploring data on a particular crime type over a number of years can tell a different story – about who is committing crimes, and how, where and when; as well as the factors that influence criminals' decision to offend.

Trends in crime are, like other trends in other areas, shaped by counterforces and are responsive to crime prevention efforts. Identifying both criminal methodologies and efforts at counteracting them is a good way of identifying who is involved in a problem situation, and why.

Where? Here?

WHY: When a problem is place-based, site visits are an essential starting point. First-hand observation from a site visit can provide rich contextual understanding of who uses the space, how they interact with it, what is lacking from (or abundant in) its built form, and so on.

"There is no logic that can be superimposed on the city; people make it, and it is to them, not buildings, that we must fit our plans."
— Jane Jacobs, *Downtown is for People*

HOW: There are innumerable methods for discerning and recording observations at a site visit. Bring a camera (see 'Point and Shoot', p.147) and try the below:

A Day in the Life — Find a comfortable place like an outdoor café, park bench, or a wall to lean on, and do some people-watching. Do people moving quickly through the space, or do they stop and linger? How many people do you see in the course of an hour or a day? Take note of the demographics of visitors to the place, but also consider how else you might describe them. Are they alone, or with someone? Why are they there? (Ask them to find out). Take photos of the space, or draw diagrams to remind you of what you saw (see 'Picture This', p.161). Afterwards, reflect on what you observed. What patterns do you notice? What further questions do you have?

Getting to know an area really well can take time and your observations will likely be different over the course of a day, week and year. An initial site visit will tell you the who, what and why of a location, but will probably raise more questions than it answers (which is a good thing). Develop field tools to answer those questions on subsequent visits.

What do you know?

WHY: The latest published knowledge will tell you what others know, and can help you contextualise what you have learned from other research methods.

"The more that you read, the more things you will know. The more that you learn, the more places you'll go." — Dr. Seuss, *I Can Read With My Eyes Shut!*

HOW: This method involves reading across a topic, reflecting on what you've read and then writing up your findings.

An environmental scan is the quickest version of this, while a literature review is typically a more in-depth piece of work and will take longer.

Environmental scan – Identify key pieces of research (journal articles, conference papers, industry papers, &c.), list them and briefly summarise their findings. Mine the references from popular publications to help you locate relevant research. Conclude with a discussion section that identifies patterns among the research you have read.

Literature review – Explore themes that are relevant to the problem within a body of work/ discipline area. Identify and meticulously record search terms, set a target number of citations based on the time you have, identify important authors and publications, and read. Take note of emerging theories and hypotheses, and what methods authors have used to gather information. Write up your findings to include an introduction, discussion and conclusion. Share it and get feedback from peers.

Futuring

WHY: Building a model, process map or a prototype of a solution can be useful to test the solution against the original problem (are we on track to solving the problem?), as an object of discussion (what do you think about this solution?) and as a way of progressing the concept from a solution idea to a prototype that has been refined over a number of iterations.

A conceptual model can also be a terrific discussion object to show stakeholders a propositional 'possible future' and give them the opportunity to discuss, critique and further develop the proposed solution.

HOW: A model can be intricate or simple. Below are some simple ways of modelling solutions:

A process matrix – On the x axis, identify who will be involved (organisations, teams, groups, individuals). On the y axis identify what actions they will take. Then consider the relationships and interactions between people and activities.

A story board – Map out a day (or week, or year) in the life of someone who you intend to use your proposed solution, and how, where and when they will interact with it.

A visualisation – develop sketches, a rendering, a virtual environment or a physical model to show the solution in place and how it will be used.

Once your model is complete, facilitate a 360° discussion with stakeholders (including users). Question the assumptions it was built on, its pros and cons, the intended and unintended consequences of its implementation, and so on, to arrive at a comprehensive overview of its strengths and weaknesses.

Picture this

WHY: Mapping and visualising is a swift and effective way to crystallise and communicate knowledge, processes, concepts, relationships, experiences, problems and opportunities.

As with role-playing or writing a narrative, showing ideas in pictures reveals things you didn't realise that you couldn't see. A talented visual communicator can interpret and represent a situation, process or problem in a way that is easy to comprehend and memorable, but also brings new insight by visualising connections, causality and flow.

However, you don't need to possess masses of drawing talent to make this method useful. In workshops, we use symbols, shapes and stick figures alongside (or instead of) words and stories. Images lodge easily and permanently in people's minds and more importantly, the process of visualising gets people talking.

HOW: Obtain a large drawing surface – typically a whiteboard or papered wall – and pens. A stick and a large expanse of firm beach-sand will also do nicely. Start somewhere and keep going – across, up, down.

Aim to be thoughtful and comprehensive rather than tidy. Question the location of and connections between the ideas, objects, symbols you are drawing and the information they represent to make the visualisation as meaningful as possible.

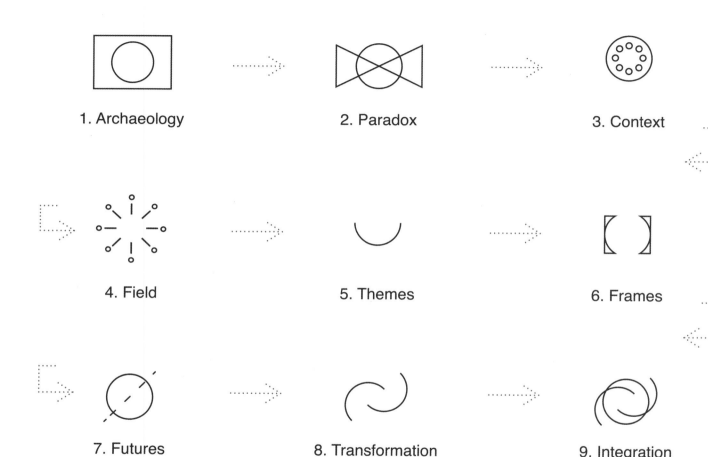

1. Archaeology

2. Paradox

3. Context

4. Field

5. Themes

6. Frames

7. Futures

8. Transformation

9. Integration

Frame creation

The frame creation workshop is a definitive moment in our process. This workshop provides an opportunity for people from the stakeholder network to come together and think afresh, using their combined knowledge; to take a long walk around the problem, suspend judgement about how to solve it, and create frames that provides new ideas. The workshop also helps generate the momentum that is required in the long term, to carry solutions to fruition.

Several versions of the frame creation workshop have been developed, ranging from a rapid two-hour version, to a five-day-long 'deep dive'. The latter leads to better results, but for the purposes of this book, we offer here the one-day version of the workshop. The methods that follow should be completed in the sequence they are presented, since each tool builds on the previous one. The first two methods are done with the 'plenary' group; if there are more than, say, eight workshop participants, later methods are done in smaller sub-groups, while others are done in pairs.

When planning a frame creation workshop, foundational research into the problem context is important to identify potential participants (stakeholders) and provide sufficient grounding in the subject matter to ensure that facilitators feel comfortable leading a group of people that will include content experts. (As a side note, facilitators do not need to become experts in content to run the workshop effectively.) Irrespective of the total number of participants, try and get as broad a spread of stakeholders as you can.

The methods shown here have been developed by Designing Out Crime and tested on many 'real' projects as well as in 'lab'-style conditions. Some have been inspired the work of collaborators from our network, and all of them have been refined through practice.

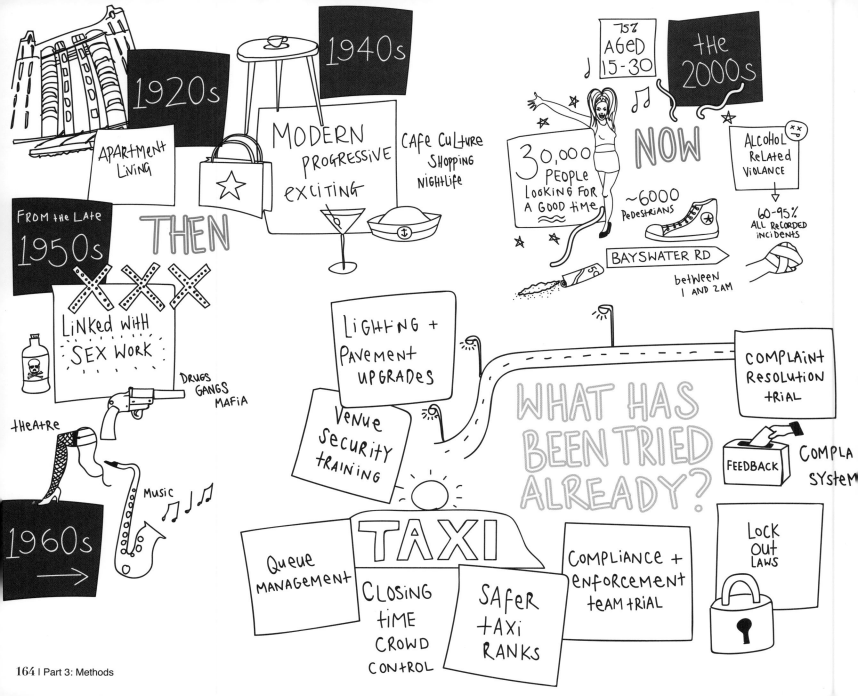

1920s

APARTMENT LIVING

1940s

MODERN
PROGRESSIVE
exciting

CAFE CULTURE
SHOPPING
NIGHTLIFE

75% AGED 15-30

the 2000s

30,000 PEOPLE LOOKING FOR A GOOD TIME

NOW

ALCOHOL RELATED VIOLANCE

60-95% ALL RECORDED INCIDENTS

~6000 PEDESTRIANS

BAYSWATER RD

between 1 AND 2AM

FROM the LATE
1950s

THEN

LINKED WITH
SEX WORK

DRUGS
GANGS
MAFIA

theatre

Music

1960s

LIGHTING +
PAVEMENT
UPGRADES

VENUE
SECURITY
TRAINING

WHAT HAS
BEEN TRIED
ALREADY?

COMPLAINT
RESOLUTION
trial

FEEDBACK

COMPLA
SYSTEM

TAXI

Queue
MANAGEMENT

CLOSING
TIME
CROWD
CONTROL

SAFER
TAXI
RANKS

COMPLIANCE +
ENFORCEMENT
teAM trial

LOCK
OUT
LAWS

1. Archaeology

2. Paradox

3. Context

4. Field

5. Themes

6. Frames

7. Futures

8. Transformation

9. Integration

Frame creation:

Digging

Organisations operate at several levels, and problems manifest in many ways. Consequently, problem-solving efforts can take many shapes and forms. They may be political, strategic, or tactical; they may involve few people or many; and they will vary in success and impact.

This tool – the first in the frame creation workshop – is used to map out all and any past actions that could be considered as attempts to solve the problem as it presents. The aim is to create an exhaustive list that provides background, but also to allow participants to 'purge' or park their thoughts about the problem, and not to fixate on what has already been attempted. This exercise can be done with all workshop participants in a 'plenary' session, or in smaller sub-groups of participants.

Duration: 20 minutes to list, 10 minutes to discuss.

QUESTIONS: What has been tried before, and to what extent have these efforts been successful?

ACTIONS: Ask the question and list the responses on a whiteboard or flipcharts. Encourage everyone to participate – there are no wrong answers – but gently dissuade participants from debating the merits or pitfalls of problem-solving efforts while the list is being made. When there is a natural pause in the room, read over the list and give participants the opportunity to comment or discuss.

LOts of venues to visit

please help

HOme to the city's poorest

A PLACe of eXtReMe WeALtH

WHAT MAKES THIS PROBLEM HARD TO SOLVE?

Not enough Public space

Reputation for eDGiness/ fringe Dwelling

Attracts 'tHe MASSes'

A LONG-STANDiNG NiGHtspot

AustRALiA's MOSt DENSeLY POPuLAteD ResiDENTIAL AReA

1. Archaeology

2. Paradox

3. Context

4. Field

5. Themes

6. Frames

7. Futures

8. Transformation

9. Integration

Frame creation:

Clarifying

In a complex problem, there are usually many pairs of oppositional forces that confound attempts to solve it. In the frame creation workshop we refer to these oppositional forces as 'paradoxes'. Identifying the paradoxes, then setting them aside, is an important early step in the workshop process and is done with the plenary group of participants.

Duration: 15 minutes

QUESTIONS: What makes this problem so hard to solve? What are the opposing forces in the problem context?

ACTIONS: One facilitator asks the questions, while another writes the responses on a whiteboard or flip-chart. Responses to the first question (what makes this hard to solve?) should be noted down and discussed, and when the responses begin to flow, facilitators can guide participants to reformulate their answers into statements of opposing forces.

Some examples of these pairs of opposing forces are described below using the example of Kings Cross in the project 'Growing Up In Public':

For a place to be vibrant, there needs to be a critical mass of people

vs.

If a place is vibrant and popular, and attracts large numbers of people, it can become unpredictable and hard to manage

and

Young people get in trouble when they go to Kings Cross

vs.

Kings Cross's reputation for edginess and danger makes it attractive to young people

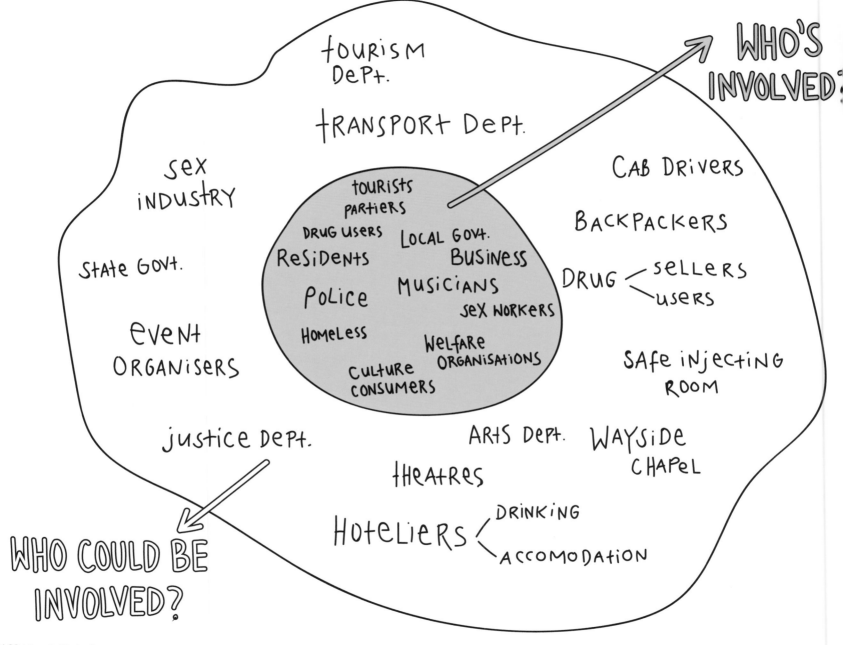

WHO'S INVOLVED?

WHO COULD BE INVOLVED?

tourism Dept.

TRANSPORT Dept.

sex INDUSTRY

state Govt.

event ORGANISERS

justice Dept.

tourists
PARTIERS
DRUG USERS
RESIDENTS
POLICE
HOMELESS
LOCAL GOVt.
BUSINESS
MUSICIANS
SEX WORKERS
WELFARE ORGANISATIONS
CULTURE CONSUMERS

CAB DRIVERS

BACKPACKERS

DRUG — SELLERS / USERS

SAFE iNJECTING ROOM

ARTS Dept. WAYSIDE CHAPEL

tHeatres

HOTELIERS — DRINKING / ACCOMODATION

1. Archaeology

2. Paradox

3. Context

4. Field

5. Themes

6. Frames

7. Futures

8. Transformation

9. Integration

Frame creation:

Expanding

People are accustomed to seeing thing through their own individual or organisational lens. This necessarily narrow viewpoint is important – it helps us get on with lives and jobs, day to day. But an essential part of exploring a problem through frame creation is expansion – making a situation more complex, adding rather than subtracting.

In this exercise, we aim to widen our view of who is involved in a problem situation. We use a specific terminology to categorise stakeholders according to how 'close' they are to a problem. In the frame creation workshop, stakeholders who are closely connected to a problem – having either been affected by it, or involved in previous attempts at solutions – are described as the 'context' stakeholders. The wider group of stakeholders – those who could conceivably be involved in a future solution – are those in the 'field'. Note that stakeholders can be specific individuals, generic groups or organisations.

This exercise – and several that follow – is generally done in sub-groups of around six people, since participants need to be able to gather around a shared writing/reading surface (whiteboard, paper, wall) and discuss what is written on it. Allow for one facilitator per group.

Duration: 10 minutes.

QUESTIONS:
Context: Which stakeholders have been a) directly impacted by the issue in question, or b) had direct influence in this problem situation?
Field: Which stakeholders could be involved in, or interested in a solution?

ACTIONS: This is a generative exercise – more is better. Write the 'context' stakeholders in the centre, and the 'field' stakeholders in the outer circle (as shown at left).

STAKEHOLDERS

WHAT'S IMPORTANT TO THEM

CITY OF SYDNEY ——————> ECONOMIC GROWTH   SAFETY   VIBRANCY

VENUE OWNERS ——————> PROFIT MARKET SHARE SUSTAINABILITY

PATRONS ——————> ACCESSIBILITY   FOOD EXPERIENCE   FUN BEING SEEN   DIVER
BLOWING OFF STEAM   EDGINESS   FRIENDS   GETTING LOOSE

BARS ——————> HAPPY MEMBERS   STABILITY   PREDICTABILITY

TRANSPORT DEPT. ——————> KPIS   RUNNING TO SCHEDULE CUSTOMER SAFETY

TOURISM DEPT. ——————> INTERNATIONAL + LOCAL TOURISM   HAVING A GOOD PRODUCT

RESIDENTS ——————> SAFETY SECURITY   NOISE   LIKING WHERE THEY LIVE

MUSIC INDUSTRY ——————> PERFORMANCE   REGULARITY AUDIENCE

SEX SHOPS ——————> CUSTOMERS   EDGINESS

*WRITE EACH
COMPONENT O
SEPARATE
STICKY NOTE

BACKPACKER
SERVICE PROVIDERS ——————> VIBRANCY AFFORDABILITY CUSTOMERS
PROXIMITY TO OTHER BUSINESSES

POLICE ——————> SAFETY   ENFORCEMENT SUPPORTED BY THE LAW

RETAILERS ——————> CUSTOMERS VIBRANCY IMAGE

1. Archaeology

2. Paradox

3. Context

4. Field

5. Themes

6. Frames

7. Futures

8. Transformation

9. Integration

Frame creation:

Thinking

By analysing what our stakeholders value, we can better understand the complexities of the problem context. This exercise is done in the same small sub-group as the previous tool ('Expanding').

Duration: 30 minutes

Questions: What is important to the stakeholders? (and creative variations, e.g. 'What gets X out of bed in the morning?')

Actions: Have the participants take a selection of the stakeholders identified in the previous tool (eight, or more if time permits). Write the name of each stakeholder down the left-hand side of a piece of paper or whiteboard and ask participants 'What is important to them?' Think about each stakeholder in succession, and have participants write their responses next to the stakeholder. (We recommend writing each response on a separate sticky note as this will be used as data in the next method.)

Encourage participants to think, and answer, in terms of 'human' (rather than organisational or bureaucratic) values or concerns. For example, if the stakeholder were a company, one response to the question 'What is important to the company?' would be 'money'. We can expand this idea into more empathetic concepts by asking why money is important, and to think in terms of the employees (organisations don't really have 'values', people do). Money is important because it gives a company a future, affords a lifestyle for individuals, is a reward for hard work… and so on. There are no wrong answers but resist the temptation to be sarcastic – empathy is one of the goals in this exercise. Aim for 'positive' values. Note that there will be much repetition, which is an inherent consequence of the exercise and, indeed, an important point to hammer home: different people value the same basic things in spite of appearances, and this realisation is crucial to designing something that will satisfy the majority.

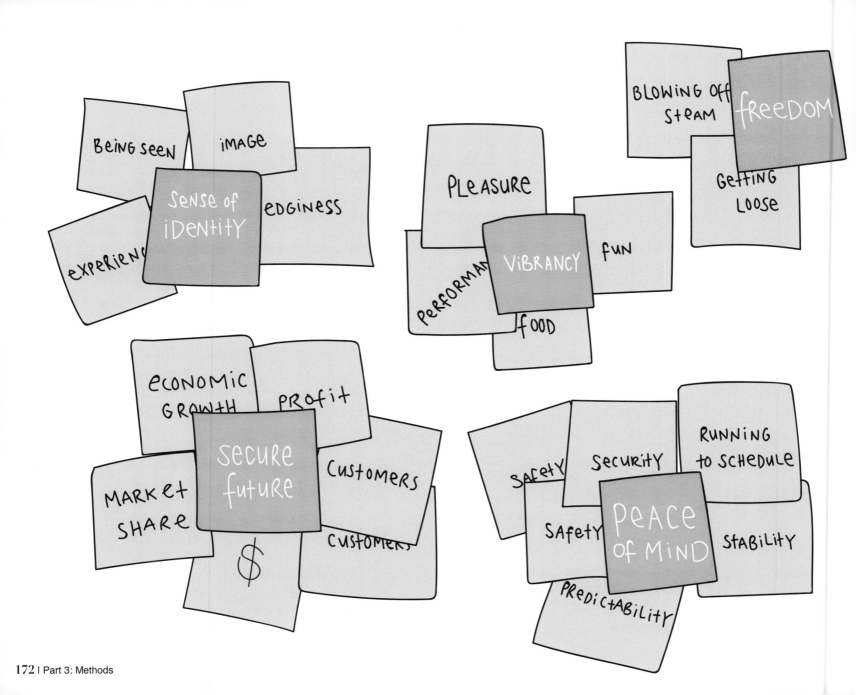

1. Archaeology

2. Paradox

3. Context

4. Field

5. Themes

6. Frames

7. Futures

8. Transformation

9. Integration

Frame creation:

Clustering

In this part of the frame creation workshop we move away from concrete analysis of the situation and into abstraction. The exercise helps participants to form an understanding of the shared interests or values of the stakeholders connected to the problem. These shared values we call 'themes', and arriving at themes is the goal of this exercise.

Duration: 20 minutes

ACTIONS: As a group, and on a separate worksheet, cluster the responses from the previous method (which are, ideally, written on sticky notes) into groups of similar concepts. Encourage everyone to participate. Remember that there will be repetition among the responses in the previous exercise, so some clusters may contain virtually a single idea, written many times.

Do this roughly and rapidly until every response from the previous method has been put into a cluster. Once this is done, discuss and analyse each cluster, and regroup clusters if necessary, by asking questions like 'What is this cluster about?' and 'Are these concepts really related, or does this idea perhaps belong in another cluster?'

Next, give each cluster a definitive name. Ask participants to drill down to the deep, shared human needs represented in each cluster. For example, of a group of words relating to financial security, ask 'what core needs drive the desire for financial security? What does financial security give us?' Personal experience can guide this exercise. Explore synonyms until the group settles on a satisfactory word or phrase that sums cluster up (and note that, as a result of this step you may need to shuffle the clusters around yet again).

Finally, place a name in the centre of each cluster. From this point in the workshop, each cluster name is now referred to as a 'theme'.

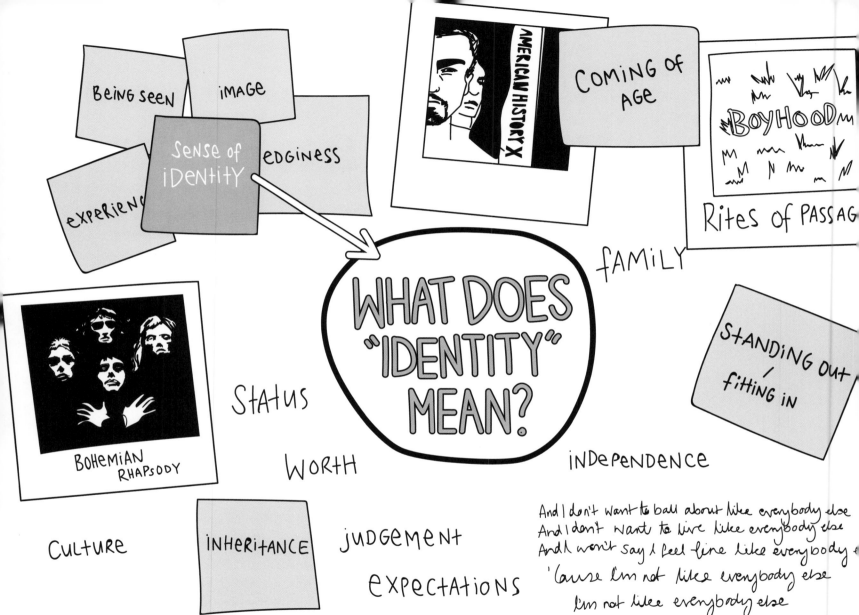

BEING SEEN

IMAGE

Sense of IDENTITY

...EDGINESS

EXPERIENC...

AMERICAN HISTORY X

COMING OF AGE

BOYHOOD

Rites of PASSAG...

FAMILY

WHAT DOES "IDENTITY" MEAN?

STATUS

BOHEMIAN RHAPSODY

WORTH

STANDING OUT / fitting in

INDEPENDENCE

CULTURE

INHERITANCE

JUDGEMENT

EXPECTATIONS

And I don't want to ball about like everybody else
And I don't want to live like everybody else
And I won't say I feel fine like everybody...
'Cause I'm not like everybody else
I'm not like everybody else

the KiNKS

1. Archaeology

2. Paradox

3. Context

4. Field

5. Themes

6. Frames

7. Futures

8. Transformation

9. Integration

Frame creation:

Analysing

Poets, authors, songwriters, visual artists and moviemakers are adept at communicating thematic meaning, showing us how themes play out on a human level. We can also understand themes by exploring experiences from our own lives. This method asks participants to analyse deeply the themes that they identified in the previous exercise.

Duration: 30 minutes

QUESTIONS: Describe a personal experience of the (theme). What novels, songs, movies, etc. convey the meaning of the theme, as you understand it?

ACTIONS: Ask participants to form pairs. Depending on the size of each subgroup and the number of pairs and themes, each pair will discuss and analyse one or more themes. Use the questions above as prompts.

To explore the theme 'sense of identity', for example, participants can think of how people (themselves, others) acquire a sense of identity or a sense of self.

Participants can ask their partners 'What identity-forming experiences have you had? When do you feel most like yourself? When and where do you feel able to express your identity?'

Personal stories are interesting and valuable, as everyone's experience will be different. The pairs then share and discuss their findings with their subgroup.

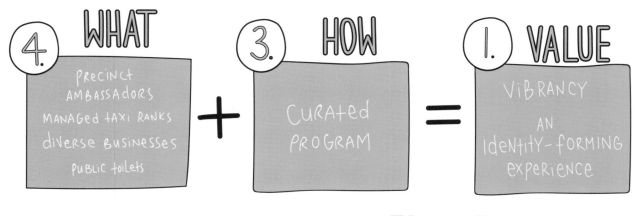

4. WHAT

PRECINCT AMBASSADORS
MANAGED tAXI RANKS
diverse BUSINESSES
PUBLIC toilets

+

3. HOW

CURAtED
PROGRAM

=

1. VALUE

ViBRANCY

AN
ideNtitY-FORMING
experience

2. FRAME

Music festival

if We see the situAtioN As oNe of **A music festival**

ANd We AdoPt its WORKING PRiNciPLe **eveNt MANAGEMENt**

tHeN

We WiLL CReAte AN eNviRoNMeNt tHAt is **ViBRANt**

ANd HeLPs YouNG PeoPLe exPLoRe theiR **ideNtitY**

1. Archaeology

2. Paradox

3. Context

4. Field

5. Themes

6. Frames

7. Futures

8. Transformation

9. Integration

Frame creation:

Framing

A 'frame' is a viewpoint that gives a different perspective on a situation. There are many ways of constructing frames and many ways in which a frame can be overlaid onto a problem context. In a workshop setting, metaphors and analogies make effective frames. To use, again, the example from 'Growing Up in Public': in exploring the theme 'sense of identity', we asked the question 'what other experiences give young people the opportunity to find/develop/express their identity in a positive way?' One answer was 'a music festival'. We then mapped out the main characteristics and operating principles of the 'ideal' music festival (i.e. the music festival we would aspire to create), and found in this context both an approach (event management) and many ideas that could be transposed from the music festival environment to Kings Cross at night.

As a rule of thumb, a 'fruitful' frame is one that makes sense to all the stakeholders and helps them envision a new approach to the problem.

Duration: 1 hour

QUESTIONS: Consider one or more of the themes (values) you have identified in previous methods. Is there a discipline area or profession that is especially known for creating these values? Is there another context or situation where these values are very keenly expressed?

Next, examine the working principles used in these alternative disciplines or situations. What activities do actors in these contexts do that could be adapted to the current problem?

ACTIONS: In subgroups, create alternative frames. Test each for fruitfulness: if a frame is not fruitful (e.g. if discussion stagnates, if no new ideas emerge) discard it and try another, or start the exercise again.

OLD

PROBLEM FRAME
ALCOHOL RELATED CRIME
SOLUTION FRAME
CRIME PREVENTION

NEW

PROBLEM FRAME
| A MUSIC FESTIVAL |
SOLUTION FRAME
| EVENT MANAGEMENT |

OLD		NEW
MORE POLICE / SECURITY "DRUNK TANKS" TOUGHER LAWS	**SOLUTIONS**	CROWD CONTROL MANAGED TAXI RANKS ③ PUBLIC TOILETS DIVERSE BUSINESSES
CRIME PREVENTION PARTNERSHIP	**SCENARIOS**	CITY COUNCIL AS EVENT MANAGER ②
REDUCE OPPORTUNITIES FOR CRIME	**GOALS**	CREATE A THRIVING LATE NIGHT ECONOMY ①
SECURITY, SAFETY	**VALUES**	VIBRANCY IDENTITY

Adapted from van der Bijl-Brouwer and Dorst, 2014, *NADI-model*

1. Archaeology

2. Paradox

3. Context

4. Field

5. Themes

6. Frames

7. Futures

8. Transformation

9. Integration

Frame creation:

Exploring

In the previous method, multiple frames were created. In this method, we aim to articulate the fruitful frames in more depth and extrapolate the new approach, goals and scenarios that arise from these new frames. We also use this method as a way of seeing the current ('old') approach in a critical light, and to understand how existing solutions are underpinned by a particular viewpoint (as shown at left).

For instance, from the new frame of the music festival, and the new approach of event management, the City of Sydney reformulated their original goal of trying to prevent alcohol related violence into the creation of a thriving late night economy. From this new goal, new scenarios and solutions could be expressed.

Duration: 5 minutes to explore each frame, 20 minutes to discuss and decide which solutions to develop further.

QUESTIONS: What organisational goals should be set to achieve the desired values? What scenarios need to be created to enable this (i.e. what roles can different stakeholders play)? What actions will need to be taken to achieve the solutions arising from the frame?

ACTIONS: Use the diagram on the facing page as a template. Start with the most fruitful frame from the previous method, and populate headings 1-3 with the data from the previous method. As an optional extra step, forensically analyse current ('old') solutions to understand the scenarios, goals and values that underpin them.

Research
Gathering the background knowledge required to tackle
the problem situation at hand

↓

Initiation
Contacting key stakeholders (individually) and interviewing them,
same for the parties in the broader field.

↓

Frame creation
Bringing all parties together in the 9-step
frame creation workshop (see p.162)

↓

Design exploration & business exploration
Exploring the frame proposals by mapping out the design possibilities
& exploring the value (for the stakeholders) of these design
concepts and ideas.

↓

Path to action
Mapping out the activities and the transformations needed
for realisation

↓

Handover
Handing over the results to the partner organisations for implementation.

↓

Evaluation
Evaluating results, process and the underlying methods.

Project model

The frame creation process that was introduced in Part 1 of this book, and the workshop methods presented here, are embedded within a broader project model.

The frame creation workshop generally forms a small part of a much longer project. But for this step to be successful, the input needs to be prepared carefully. This can be a time-consuming process, often taking several weeks or months.

Following the frame creation workshop, the new frames and resulting solution directions need to be explored, creatively and critically, before the initiative can be handed back to the problem owner/s.

4.

Mov

ng
ahead

Inspiration

We take great inspiration from sharing our experiences, and in being told by our peers and partners that designing for the common good is both necessary and quite possible. We have tried to convey some of the tricks of the trade, as we have discovered them over the last seven years, and hope that this is valuable to you.

What remains harder to express are the personal rewards of this work: the joy of seeing people unlock 'impossible' problem situations and get to real solutions; the struggles when unforeseen obstacles crop up; the frustration of realising that in your enthusiasm you ignored early warning signs; and the sighs when you have to retrace your steps and resume the project from the junction where earlier you took the wrong track.

In spite of the many challenges involved, there is something addictive about designing for the common good; it becomes a compulsion. Ignoring one's ability to make a real contribution to society comes to feel like betraying not only the other, or the good cause, but also oneself.

The three personal stories on these pages tell the personal tales of a designer, a criminologist and an artist who are pioneers and leaders in this field. They show how three very different people with very different backgrounds have found that they can use this design-based approach to contribute to the world.

Douglas' story

Douglas Tomkin studied industrial design at Royal Melbourne Institute of Technology in the 1960s – at a time when design students were heavily influenced by authors like Victor Papanek (*Design for the Real World*), who inspired designers to take responsibility for the effects of mass production and consumerism and exhorted them change the world for the better through design. Upon graduating he began designing special furniture for the elderly. This was a type of designing that nobody had really looked at before: design ergonomics was an emerging field of research into the dimensions of the human body, its perceptual and cognitive abilities, capacity for movement and so on. It was clear to the young Douglas that this type of research and its resulting knowledge would be essential in designing quality products for vulnerable groups, and for all of us.

When he moved to London, he worked at the Royal College of Arts as a research assistant to Professor Bruce Archer, one of the very first researchers in design. The group at the RCA pushed the boundaries of design for the public service through projects such as the development of a revolutionary, adjustable hospital bed. Douglas moved into design education and became head of the school of design at Hong Kong Polytechnic, where he built a design curriculum for the Asian context that included a solid grounding in ergonomics and methodology. Moving back to Australia he transformed the Industrial Design course at the University of Technology Sydney from a commercially-oriented degree to one more deeply engaged with society ergonomics and sustainability. He also worked as a forensic designer, using his specialist

Hope

The projects we have presented in this book are often seen as part of a much broader movement called 'social design'. While this is true in a general sense, we also like to keep our distance from that category a little bit. In contrast to a lot of 'Social design' projects we take a more bottom-up approach, and aim to achieve real, 'on the ground' impact – all of our projects are situated in reality. This inevitably limits their scope, but that need not limit their impact at all, especially in the long run. We work with society's problem owners to create new ways of addressing the issues they face. We hold that progress in furthering the common good in our societies will come from a multitude of such precise, targeted and concrete project interventions. And we have seen that the principles and patterns that emerge in these projects can then be applied more broadly, in other, comparable situations, and thus eventually add up to a comprehensive solution.

This does, however, mean we need many of these interventions to make a real impact. That was the reason to write this book now. We hope to inspire you to pick up this knowledge and run with it – in your own way, in your own situation. There is so much that needs to be done and this is only the beginning. Let's start a movement.

expertise in court to trace back the origins of accidents and mishaps either to user carelessness or design faults.

Over the years, the focus of his design practice shifted from ergonomics towards a more interdiscplinary approach, where he worked directly with specialists in other disciplines. He found that inter/multidisciplinarity offered a different way to achieve high-quality design.

Joining the Designing Out Crime team as the only designer among criminologists, architects, a psychologist, historian and a philosopher sparked yet another shift. In this environment design was no longer the dominant profession, as the strength of these projects was built on the richness of multidisciplinary interdependence. In such a situation (of complex creative processes, dealing with the most complex social situations and issues), design cannot be the answer, nor does it need to possess the solution.

Rodger's story

Rodger studied psychology in the 1990s. Having grown up in a family plagued by mental illness, he wanted to be able to help people who needed it. After university he joined the public service, providing support to victims of crime in a team that took applications for counselling and made referrals. He found the job of helping people through difficult times both taxing and deeply rewarding.

But these were victims of crimes, some of which could have been prevented, so he moved to the crime prevention area of the department, where he worked with other government agencies, local councils and community groups on projects

Momentum

Luckily, we are not alone: projects in which design practices are applied to social problems for the common good are happening all over the world. We have worked with colleagues in Denmark Indonesia, Holland, China, South America, South Korea, and the UK, and we are discovering new initiatives every day. Our approach may be slightly different because the practices presented in this book are based on many years of research that teases out the deeper elements of design. In so doing, we have moved away from what designers in the field would see as 'normal' design practice, and have adapted designing to this new and very complex field. Designers reading this book will have realised by now that designing for the common good does require some approaches and techniques which, although known in the design professions, are not very common.

After traveling through theory, we have landed again in the real world and with our partner organisations have worked towards establishing a proof of concept by building up a portfolio of over 140 projects, only a fraction of which could be presented in this book. Some of our earlier projects were quite a struggle, as even our most willing partner organisations had trouble understanding how the approach would work, and seeing the benefits it could bring. Lacking proof or examples, partners had to be convinced on the basis of ideas, which is always difficult. We have to thank these partners, in particular the Department of Justice, for their trust and patience in those early days. Now that so many projects have been done, the burden of convincing people is much lighter: this book is full of examples that demonstrate that the design approach to social problems works.

aimed at stopping crime. During this time he completed a Masters degree in criminology.

After a few years of working in this role he became frustrated with the discipline of criminology, and how it was being applied in public policy and practice. The insistence on evidence-based interventions meant that very pertinent issues were not being addressed, since there was no evidence base – no best-practice guide – for tackling them. Meanwhile, 'evidence-based' solutions weren't necessarily working, and the department had no mechanism for coming up with new or novel approaches. It felt like tinkering at the edges while problems festered and grew.

In 2010, an opportunity arose for a secondment at the Designing Out Crime research centre. The centre was at that time still a relatively new initiative of the Department of Justice and the University of Technology Sydney, having been set up to develop an approach to crime prevention that derives from the discipline of design (among others) rather than criminology. The collaborative and explorative way in which the Designing Out Crime team worked with its partners – investigating problems and possible solutions together with organisations rather than dispensing pre-packaged solutions to them – was new and refreshing.

After a couple of years working together the Designing Out Crime team took stock of the projects, and started articulating the tools and methods that had worked in reaching new outcomes. Rodger enthusiastically took the lead in creating a pack of method cards, drawn from each of the disciplines (architecture, criminology, design, history, psychology, urban planning), some of which are included in this book.

Movement

One of the great joys of these projects is that they provide an opportunity to gather very different people together and offer them a framework for achieving progress on a very stuck problem (indeed, this is the main aim in a project like 'Teaming Up', p.78). The best projects by far are the ones in which people from different professions and organisations are personally inspired to bring their organisations along for the journey. Then the frame creation process has the potential to create a groundswell movement that will lead to lasting change in a whole sector, far beyond the horizon of the single project.

The role of the facilitator then shifts to supporting these 'champions' through a process of often quite disruptive organisational change. Radical change is inevitable: if we look at an organisation as a kind of problem-solving machine, very finely attuned to performing its routine tasks and solving the problems associated with them in a set way, then a new frame that entails a new approach to these problems inevitably requires dismantling established processes and organisational structures. With this book we hope to have empowered you to become such a champion.

We are convinced that in the long run, organisations will see the validity and relevance of this new approach to the problems they are facing. But saying that in the long run 'resistance is futile' doesn't help us now in getting organisations to change. In the next steps of the development of the designing for the common good story, we will need to move beyond the doing of wonderful and elegant projects towards addressing these organisational issues, both directly and indirectly.

Rodger is now a PhD candidate, researching how the designing out crime methodology can help people tackle complex problems in a way that is targeted, but open, and lead to solutions that can have a real impact on people's lives.

Peik's story

Peik Suyling is a Dutch artist and designer, and a long-standing pioneer in social design. He is the director of the Young Designers foundation (YD/), which was established to initiate and develop design projects within the context of cultural and societal change.

The YD/ foundation has always sought to be completely free and unflinchingly open in its approach: each project is called a 'quest', and is shaped by deep understanding of the content of the problem. Authenticity and conceptual beauty, as defined and redefined in every project, keeps the designers and stakeholder aligned through what is, for many of them, quite an adventurous and scary process. The reward is a profound and often life-changing learning experience.

YD/ was initially supported through cultural funds, giving the foundation a position of independence and freedom to experiment and an ability to be uncompromising in its commitment to quality and depth of learning. However, after the Global Financial Crisis, this cultural funding dried up. In this situation Peik decided to continue 'design for society' activities from the position of an autonomous artist. This meant projects would come no longer from external

Next steps

Over the years the focus has shifted from working on projects – which are great for delivering a proof-of-concept but might be unable to deliver lasting change – to delivering programs to partner organisations. These programs are (in our case) three-to five-year contracts that can involve projects, consultancy, masterclasses, staff training and whatever is needed to help our partners adopt a design-based approach to problem solving.

Signing up to such a program is quite a step for an organisation, as it is aimed at changing the organisation itself (rather than something 'outside'). This is innovation of the more painful kind, and requires top-management backing and preferably top-management participation. Masterclasses, short projects and other experiential formats then help to engage a wider circle of staff. The fact that the Designing Out Crime research centre and its big sister, the Design Innovation research centre, are based at a university has allowed the development of the taught elements of these interventions into a curriculum for employees from the partner organisations. Such capacity building is very important. The three personal stories you have read present three people with very different histories, each bringing their own approaches to designing for the common good. These exceptional people are strongly committed for very personal reasons. Unfortunately, they are not only exceptional, but also exceptions. An organisation wanting to change cannot generally sit around and wait for such people to knock at the door. It needs to inspire its current people to drive the change.

One of the initiatives we have developed, especially for the public service, is aimed at creating an entire generation of public servants for whom a design-based problem-solving approach is second nature.

opportunities, but would have to be initiated by Peik himself. This led to a new period of soul-searching, in a very literal sense of the word.

At one point he realised that from a young age he had dreamt of becoming a baker. And so, he bought a trailer, hand-built a clay oven inside, and began baking bread. He called his mobile bakery 'de Eenvoud' – Dutch for 'simplicity' or 'artlessness' – to express the purity of its purpose and intention. He had wanted to find a way connect with others at a very deep, basic level, and felt that by staying close to the core of his own identity, he could achieve these kinds of relationships.

Now, he goes into neigbourhoods, bakes bread and invites people into the trailer to sit at the table next to the warm oven, smell the bread and talk. Some people stop and want to buy the bread. They are invited to come in and bake their own, and have a cup of coffee while they are waiting. The conversations are wonderful. People immediately recognise the bakery as the articulation of something quite existential, and the discussions and silences naturally gravitate to that existential level. Bread is something we all have in common, part of the long-term-memory of humanity, and the bakery helps to articulate and give people a feeling of the simplicity of life.

Peik's experience is that real change only happens when people get back the core of their personhood; making changes on the surface of life just doesn't bring any sustained change. From this realisation, there is no turning back.

Next gen

A good number of the projects showcased in this book were done in partnership with public sector organisations. Through the relationships we have nurtured, we have realised that there is a growing appetite for innovation and creativity in the public service, that reflects a public expectation that government should respond to community need in an increasingly quick, agile and effective way.

Many diversely skilled and gifted people choose to dedicate their careers to the public service, rather than using their talents in the private sector. Collectively, public servants have the ability to be the creative engine for an innovative government, but individuals working within the public service often feel stifled and crushed by what appears to be a preoccupation with the way things are done, rather than the way things could be done. Many public servants' working days are consumed with explaining, justifying, proving and demonstrating, leaving little room for activities related to the creation and generation of new ideas. Sadly, wasted talent and enthusiasm begets declining morale and perpetuates a culture of stasis and business-as-usual, giving the public service an undeservedly bad reputation.

In working with public servants we have seen both a great hunger for new ideas, and considerable aptitude for adopting the type of problem-solving approach that we offer. This has led to an education program to inspire and educate the next generation of public service leaders to take a new approach to society's most complex problems. This program provides a physical and mental space for problem exploration and innovation, and encourages a free-thinking, experimental, approach that makes the most of individual talent and interest.

Watch this space.

index

authors

Kees Dorst

Kees is Professor of Design Innovation at the University of Technology, Sydney. He also holds a professorship in Entrepreneurial Design of Intelligent Systems at Eindhoven University of Technology in The Netherlands. He is founder and director of the UTS Design Innovation Research Centre and the Designing Out Crime research centre. He lectures at universities and design schools throughout the world, and has published numerous articles and books – most recently the books *Understanding Design: 175 reflections on being a designer* (2006), *Design Expertise* (2009) with Bryan Lawson and *Frame Innovation: create new thinking by design* (2015).

Lucy Kaldor

Lucy joined the Designing Out Crime team in October 2010 and is an experienced workshop facilitator and researcher. Her work has included translating and developing the frame creation model into practice and integrating it into the centre's teaching materials. Lucy leads the undergraduate teaching program at Designing Out Crime and instructs on a number of professional development programs including the NSW Public Service Commission's innovation leadership development program. Lucy holds a Bachelor of Arts with Honours in History from the University of Sydney and brings research methods and perspectives from this discipline area.

Lucy Klippan

Lucy holds a Bachelor of Fine Arts from the University of New South Wales and a Master of Design from the University of Technology Sydney. Lucy is a visual communications designer with a varied background that includes work in marketing and communications in both the private sector and the public service. Her work at Designing Out Crime spans consultancy on local government safety strategy, co-design facilitation with government agencies and other organisations, graphic design and illustration, and managing Designing Out Crime branding.

Rodger Watson

Rodger joined the Designing Out Crime research centre at the University of Technology Sydney in 2010. As Deputy Director of Designing Out Crime, he has overseen the development of its projects and programs. He holds a Bachelor of Arts (Psychology), and a Masters of Criminology and his PhD research is on problem solving in the public service.

contributors

Lindsay Asquith

Lindsay has a PhD in Architecture from Oxford Brookes University and has written a number of journal and magazine articles as well as edited a book on Vernacular Architecture in the 21st Century. She worked at Designing Out Crime for three years and was involved in a variety of projects including designing against petrol theft, design of Barangaroo Headland Park with safety in mind and mitigating the risk of armed robbery in bottle shops. She is currently Night Time City Project Manager at the City of Sydney, managing projects that increase the vitality and diversity of the city at night while also maintaining a safe and sustainable night time economy.

Mieke van der Bijl-Brouwer

Mieke is Senior Research Fellow at the Design Innovation Research Centre at the University of Technology Sydney. Her research spans the fields of human-centered design methodology and social innovation, by investigating how human-centred design methods contribute to tackling complex societal problems. For that purpose she works with public sector organisations across domains such as education, mental health, housing, crime and community. In 2012 she gained a PhD (cum laude) on a study into user-centred design methods from the University of Twente. She has published extensively in the areas of human-centred design, collaborative design, design education and social innovation.

Olga Camacho Duarte

Olga is a researcher, facilitator and consultant, with a Doctorate in Management. Through her teaching in various universities in Sydney, she is contributing to the learning and growth of the next generation of business leaders and entrepreneurs. Olga combines her background in design with her expertise in business and social sciences research to contribute to service innovation in the non-for-profit and government sectors and also works in marketing and customer experience for non-profit organisations. During her time at Designing Out Crime she led research projects that focussed on design-oriented solutions to complex social issues.

Nick Chapman

An environmental scientist by training, Nick has focussed in the last 10 years on combining local government place management practise with urban sustainability research and postgraduate teaching as a senior lecturer at the University of Technology Sydney. He is currently Place Manager for the Inner West Light Rail and GreenWay Corridor. His work with Designing Out Crime has provided students with valuable opportunities to address real-world challenges and present their solutions to the stakeholders who are tackling these challenges on a daily basis.

Rohan Lulham

Rohan is a research fellow at the Designing Out Crime research centre and holds a PhD in Environmental Psychology from the University of Sydney. His research covers the areas of design, environmental psychology and criminology, with particular research interests in affect and design, correctional design practice, and social innovation. He has expertise in social science statistics, and qualitative methods are a feature in most of his research. He is currently leading projects that seek to bring innovation to correctional design practice as well as a program of research exploring the possibilities for utilising Affect Control Theory in design research and practice.

Ilse Luyk

Ilse has a Masters degree with Honours in Industrial Engineering and Management Sciences, and in 2009 finished her PhD in the department of Technology Management (both at the Eindhoven University of Technology, the Netherlands). She was Assistant Professor at the Eindhoven University of Technology from 2007 to 2011. Her main research interest is design thinking and business innovation. In 2014 Ilse started working as product manager at De Persgroep Nederland, a Dutch media cooperation. In this job she is involved with a variety of multidisciplinary product- and business-development projects.

Dick Rijken

Dick is professor in Information Technology and Society at the The Hague University of Applied Sciences, where his research focuses on the development of teachable methods and techniques for conceptual thinking in the context of complex problems. He is also director of STEIM, a lab in Amsterdam where artists and scientists experiment with sound art and electronic live performance. His primary interest is the application of an artistic mentality to problems and issues in many different sectors of society.

Rob Ruts

Rob is an artist, focusing on urban development, and specifically on the risks that come with bringing together many people in the tight space of a city. He artistically explores how urban conflict can act as a starting point for creating new ways to develop urban spaces. This leads to projects ranging from designing police training to an art project that documents the urban knowledge of cart pullers in Mumbai. Currently Rob is a researcher at The Hague University of Applied Sciences.

Peik Suyling

Peik is a Dutch social designer, artist and director of the Young Designers foundation. The projects that he initiates with others touch on social issues. They concentrate on deep human themes like 'feeling at home', identity, independence and loneliness. Recent examples include 'de Buurtwerkplaats', a social enterprise for a neighborhood in Amsterdam West; and bakery 'de Eenvoud' (see Peik's personal story in Part 4 of this book). He teaches at various art academies and design schools.

Douglas Tomkin

Douglas is an Associate Professor in the Faculty of Design Architecture and Building at the University of Technology Sydney. He was a founding member of the Designing Out Crime research centre and is mightily impressed how a small multidisciplinary team has achieved so much in such a short time. Prior to joining Designing Out Crime he was Head of the School of Design at the University of Technology Sydney, with a strong interest in social and sustainable design issues. Douglas also acts as an expert witness on design-related court cases. Before coming to Sydney in 1992 he was chief executive of the Hong Kong Design Innovation Company.

Kim Wan

Kim holds a Master of Planning from the University of Technology Sydney and a Bachelor of Science in Business Information Technology from the University of New South Wales. Kim is a member of Designing Out Crime's consultancy staff with a background in urban planning. Kim consults on Designing Out Crime projects concerning urban contexts including transport interchanges, nightlife precincts, urban parks and public housing. She also brings experience in local government strategy and development assessment, and over 6 years consulting experience with various private and government organisations.

Vera Winthagen

Vera Winthagen is a Strategic Design Consultant for the Municipality of Eindhoven. Applying design thinking methods, she works on topics such as 'smart cities' and 'redesigning government', involving citizens in the government's decision-making process. She sees herself as a catalyst for social innovation. Previously she was project leader for Social Design and Designing Out Crime projects at VanBerlo design and the department of Industrial Design at Eindhoven University of Technology.

Jessica Wong

Jessica holds a Bachelor of Industrial Design from the University of Technology Sydney. With a wide range of skills, specialising in CAD, prototyping, and visual communication, she works with Designing Out Crime's students and clients to develop concepts into prototypes, and assists in developing the centre's range of publications and presentations.